MAPPING MANIFEST DESTINY

Chicago and the American West

An exhibition at the Newberry Library
November 3, 2007–February 16, 2008

Curated by
Michael P. Conzen
Diane Dillon

The Newberry Library
Chicago

ISBN: 978-0-911028-81-2
ISBN-10: 0-911028-81-1

CO–CURATORS: Michael P. Conzen and Diane Dillon

EXHIBITS MANAGER: Riva Feshbach

EXHIBIT DESIGNER: Chester Design Associates

CONSERVATOR: Giselle Simon

PHOTOGRAPHER: Catherine Gass

CATALOG MANAGING EDITOR: Douglas Knox

CATALOG COPYEDITOR: Sarah Fenton

CATALOG DESIGN: Kimberly Zingale, Poccuo

Printed in the United States of America by
Yorke Printe Shoppe, Inc.

Mapping Manifest Destiny: Chicago and the American West has been made possible by major underwriting from Barry and Mary Ann MacLean, with additional funding provided by Andrew McNally IV. Additional support is generously provided by Northern Trust.

The Newberry Library
60 West Walton Street
Chicago, IL 60610
www.newberry.org

Contents

Preface

The Newberry Library possesses one of the foremost collections of historical maps in the world. The depth and breadth of these collections have become especially visible in the Fall of 2007, with the Newberry's contribution of 100 maps to the remarkable Festival of Maps, Chicago, a citywide celebration of how people across the world and across the ages have created visual representations of space—real and metaphorical.

Mapping Manifest Destiny: Chicago and the American West is one of three major exhibitions the Newberry has organized or co-organized as part of the Festival. The Library lent its curatorial expertise, in the persons of James Akerman, Robert W. Karrow, Jr., and Diane Dillon, and several of its most important maps, to the exhibition mounted across town at the Field Museum, *Maps: Finding Our Place in the World.* A second exhibition at the Newberry, *Ptolemy's* Geography *and Renaissance Mapmakers*, celebrates the richness of the Library's holdings of the great works of Renaissance cartography. The project at hand, *Mapping Manifest Destiny*, exploits the depth and breadth of the Newberry's holdings of books, maps, art, and ephemera documenting the history and geography of the American West from the first Euro-American contacts to the beginning of the twenty-first century.

The history of the mapping of the American West is a compelling but complicated subject; no single work can hope to capture all of its nuances. As works of scholarship, exhibitions face particular challenges in treating a subject such as this. The format favors the display of objects and imagery at the expense of narrative text, requiring curators to elucidate big ideas in small and digestible elements. But this apparent limitation also can be liberating. Freed of the necessity of footnotes and lengthy explications, curators can focus on what seems most essential. Great exhibitions can wrestle with new ideas, suggest new directions for research, and encourage audiences to consider familiar subjects with fresh eyes. So it is with *Mapping Manifest Destiny: Chicago and the American West.* The occasion of the citywide Festival of Maps, and a suggestion from map collector Barry MacLean, provided the impetus for telling this story with an emphasis on Chicago and its cartographic industry, placing the city within the larger narrative of American economic, demographic, and political expansion into the West. Though the exhibition is roughly chronological in organization, each new section introduces a new theme and opens a new perspective on the mapping of the West, from the deployment of maps in imperial competition to the use of maps in government and nation-building, in educating the public about the West, and in promoting and facilitating commerce and economic development. Thus, as we move forward in time, the meaning of the phrase "mapping manifest destiny" broadens and deepens. With the exception of a few notable and very much appreciated loans—especially from Barry MacLean's outstanding collection—the exhibition draws almost entirely on the Newberry's collections. It draws equally on the expertise of its talented curators, Michael Conzen and Diane Dillon, who have translated an idea into a narrative, and a set of remarkable objects into an equally striking work of historical scholarship.

James R. Grossman
Vice President for Research and Education

Cities and the Mapping of Frontiers: The View from Chicago

Michael P. Conzen

The American West has for centuries attracted fascination and competition as a realm of mystery and promise, of natural wonder and human potential, of desire and challenge. For much of its human history it has served as the prototypical frontier zone on the vast periphery of ambitious colonizing societies situated in an Old World intent on occupying and transforming it. Its geographical position, scale, and diversity have compelled numerous nations and countless individuals to learn its composition and shape it according to their own designs. Inevitably, dreams of conquest set sights on the entire continent. Once the United States had formed it was not long before political ambition sought to expand its territory clear to the Pacific Ocean. A vast descriptive library developed to interpret land and life in the West, permeating everything from writing to visual arts, music, and drama. Maps played a crucial role in all this cultural production. They served as essential tools of wayfinding and subjugation, as inventories of valuable resources, and as pathways to exploitation and consumption. With their ability to fix reality in space, maps were among the most persuasive sources of information in the national project of pursuing Manifest Destiny.

Much has been written on the mapping of the American West. The nature of maps as complex artifacts of visual representation has made them attractive subjects of museum and library exhibitions. And yet, rarely has the relationship between Western maps and the cities where they were produced been explored, though maps have been quintessentially products of urban settings. This is especially true for printed maps, whose authors have sought wide audiences, and whose production required printing and allied facilities available only in large towns. As the West loomed ever larger in the American mind, those production centers at first were the East Coast cities—Philadelphia, New York, Boston—where capital was so long concentrated, for investment both in the West and in the maps depicting it. But over time the growth of internal capital helped map publishing diffuse westward itself, not to every little town with a primitive printing press, but certainly to up-and-coming new metropolises challenging the Old Order. Chicago was a classic case of a brash and successful entry into the upper echelons of a specialized form of information management, which is essentially what maps represent. By taking into account where maps were produced, this exhibit considers the role maps played in gaining mastery over a domain as vast as the American West. Chicago occupied a constantly shifting position within and beyond the "rolling frontier" of that West. By the early twentieth century, Chicago had become the undisputed national leader in the mass production and marketing of America's most widely used maps.

While the subjects and style of maps treated in exhibitions has become more diverse, the temptation has remained to favor—if not to wholly privilege—'antique' maps that are aesthetically irresistible but thematically restricted. Too often later developments have been ignored or shortchanged as mapping became more specialized in its functions, varied in sponsorship, and diverse in technological basis and graphic appearance. This Chicago-oriented exhibition of the cartography of the American West attempts a conceptual connection to the Eastern and Midwestern metropolitan centers that for so long dominated the production of American maps about, and for, the West, and which powerfully shaped the emerging national image of this vast region. It highlights the successive shifts in perspective that accompanied the changing locus of Western cartography from the sixteenth century to the early twentieth century. During the colonial period maps depicting the American West emanated almost exclusively from Europe. During the early national period East Coast centers dominated American mapmaking. After the Civil War Chicago printers and publishers rose rapidly to challenge Eastern houses, and several Chicago firms dominated the field throughout much of the twentieth century. The exhibition, its catalog, and this essay explore how the geographical complexity of the American West—both as physical environment and as human settlement space—was represented in an increasingly diverse array of maps. They also examine how mapmakers in Chicago (the new gateway to the West once its radiating railroads superseded the more limited river regime of St. Louis) became the West's most ardent and prolific delineators. In doing so these mapmakers shaped the modern cartographic image of the West, particularly in the popular mind.

In this exhibition the West is conceived as whatever territory lay west of the densely-settled and well-developed zone forming the core of the nation at any given historical moment. In colonial times this meant most of the continent beyond the Appalachian Mountains bounding the Eastern Seaboard. In the antebellum period it meant the Old Northwest and the lower Mississippi Valley and all the river's western tributaries. After mid-century it included California, western Oregon, the High Plains, Rocky Mountains, and Great Basin, meanwhile losing eastern territory to a nascent Middle West. Thanks to gold strikes, Alaska would merit a place on the national mental map of the American West by the early twentieth century. If conceived as having outer geographical limits at all, this super-region has continually shrunk, as the Appalachians, the Mississippi River, and then the 100th Meridian lost plausibility as dividing lines between "East" and "West." But in another sense the West has always been a vast realm loosely defined more by its contrasts with the nation's well-established regions than by any unifying geographical attributes within it. We make no apology for including maps showing the full United States, because the extent to which, as well as the ways in which, the West was incorporated into a culminating cartographic picture of the nation as a whole is as significant to the larger story as any map of its parts.

The exhibition also argues for a cultural understanding of maps as social documents. Geographical information becomes culturally relative and subject to interpretation in the hands of those on a mission and with particular views of the world. Maps and atlases presenting the locational arrangement of Nature's physical features such as rivers, mountains, and coastlines might in themselves seem objective enough. Even the addition of state boundaries, towns, roads and railroads, depicted as static cartographic symbols, could be regarded as ideologically neutral. But what such maps include and exclude, and how that information is represented, depends on models of reality grounded in some ideology or other. From maps setting out to be scientific and objective, to those shamelessly asserting political and ideological positions, every map speaks at some level to the world view of its maker or makers.

Four dimensions to mapping the American West

The themes of this exhibition will be traced in four sections that represent two basic types of mapping activity, and two logical subdivisions within them. The primary division is between maps created in the public and in the private spheres, although the boundaries have never been particularly sharp.

Maps produced by the state or one of its agencies clearly imply that their purpose is to serve a national or public interest, in which the effort made to create them is supported by the commonweal. There is a benefit to all citizens from such maps, which justifies their all sharing the cost. Privately produced maps follow a different logic: that special interests with specialized cartographic needs create a more limited market for their product, and that the risks attendant to supplying them should be commensurate with the possibility of a profitable reward. There have long been differences, blatant and subtle, between maps produced by these two spheres, even when the theories and conventions of mapmaking have tended towards a common vocabulary and visual grammar in map design. In the context of the American West, this primary distinction between 'government' maps and privately-produced maps has large relevance. Maps resulting from exploration, land surveying, resource appraisal, and military activity were classically public products that no private enterprise could muster the capital or manpower to produce, and yet they created and diffused knowledge that in turn spurred broad-based settlement and development. Beyond such essential work lay a plethora of optional mapping needs and opportunities that government did not have the capacity to provide; private publishers did the work instead.

The public or state-sponsored sphere is here taken to mean maps needed to advance the interests of empire- and nation-building. This divides simply into maps of the American West made before and after the emergence of a sovereign United States of America. Maps produced by colonial powers up to the eighteenth century differed in obvious ways from those made by the American Republic following independence, and so justify separate consideration. The private sphere considered in this exhibition comprises exclusively American maps made by private interests covering the West. On the private side, notwithstanding the pecuniary considerations in assessing whether to publish maps at all, is a distinction between maps offered for general enlightenment and those made to serve business, broadly defined—in other words, maps to assist in making a living.

Hence, we have maps for empire, maps for building a nation, maps for enlightenment, and maps for business. These four categories, insofar as they reflect common challenges and solutions, allow fundamental themes to be followed across time and space.

The Chicago factor

The urban component in all this, and the position of Chicago within the world of map making, shifts character with each category. It may have witnessed human traffic for many centuries, but the Chicago region in colonial times (and especially what became known as the region around the Chicago Portage) was barely a fixed spot on anyone's mental or consciously drafted map. Here the interest lies in observing how the site of the future city became more specifically identified and defined. As national expansion pushed west and the site amassed commercial promise, Chicago became a place to be mapped both in its own right and as part of the mapping deemed necessary to facilitate modern development in the area. In this context, Chicago lay in the very heart of the American West as it was perceived and as it functioned. Much of the mapping of mid-century and later was government-supported, and could therefore not accrue to Chicago, since the seat of federal government needed production nearby, giving Baltimore and New York engravers and printers the lion's share of contracts.

As Chicago turned into a wildly successful 'shock' city, growing exponentially during the second half of the nineteenth century, it surpassed the critical mass needed to attract 'venture cartographers', makers not only of utilitarian maps but also of maps that expressed visions of national growth relating to history and moral character. But it was maps that answered the needs of commerce that really spurred the rise of a Chicago map industry to rival that of long-established East Coast centers, principally Philadelphia, New York, and Boston. By this time, the American West was out on the high plains—among the Rocky Mountains, along the Pacific Rim, and even in frosty Alaska—and Chicago's immediate surroundings became simply the American 'Middle' West. For a still expanding country, the need to overcome huge distances with ground communications heavily privileged nodal points midway between coastal extremes.

But in our thinking about map production and use, the competitive 'urban card' for any particular city can be overplayed. Wherever they were produced, maps were

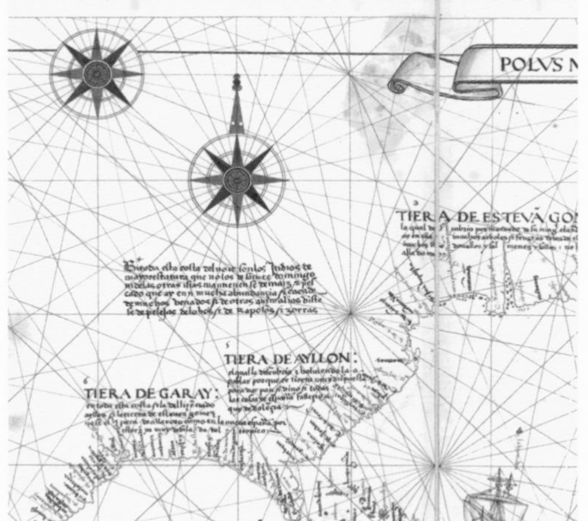

FIG. 1: Detail of Ribeiro 1529; see Item 1.1.

FIG. 2: Detail of De Fer 1718; see Item 1.13.

FIG. 3: Detail of Lahontan 1703; see Item 1.11.

light in weight and rich in information, and their value easily withstood the cost of shipping through an extensive mail and transportation system. It is not that Chicago gained an exclusive hold on Western mapping, for it did not. Rather, in an environment in which maps of and about the West were produced in numerous cities and some small towns—including in the West itself—Chicago was able, against some odds, to dislodge its big Eastern rivals (notably Philadelphia and later New York) as the premier center for certain important types of maps. Over time, maps produced in Chicago entered the national distribution stream in ever increasing numbers, and in several categories emerged as national leaders. This was not easy. Eastern cities remained competitive through the scale and sophistication of their printing and marketing facilities, as well as the concentration there of capitalists, entrepreneurs, and would-be emigrants with strong needs for cartographic knowledge about the beckoning West. But owing to strong and diverse geographical advantages, Chicago helped extend the nation's manufacturing belt as well as commercial core from the Eastern Seaboard into the Middle West, and so gained similar concentrations of technical know-how and venture capital focused on Western development.

This exhibition is organized in the four sections mentioned earlier: empire, nation-building, enlightenment, and business. The individual character of maps of the West and the sheer range of their provenance and purpose defy such neat classification—or any other 'categorillas' that might have been devised—and the viewer's indulgence is asked in recognizing that particular maps often straddle these groupings. The following overview includes reference both to maps featured in the exhibition and to some not featured.

Maps for empire

The European colonial presence in North America injected explorers, military forces, traders, administrators, and settlers onto the continent, upsetting a pattern of small populations inhabiting broad stretches of land, depending for their survival on close association with Nature's offerings and hazards, and not infrequently in conflict with each other. Maps had more far-reaching utility for the explorers and conquerers than for the indigenes, and the differential survival of maps made by each has left us with an interpretive imbalance hard to overcome. To begin with, then, the 1529 world map by Diogo Ribeiro [Fig. 1] is the earliest general map depicting North America with any semblance of verisimilitude, and opened a new era in the European understanding of the continent. Nicholas de Fer's 1718 map of the Mississippi Valley shows the advance of knowledge of the continental interior by the early eighteenth century [Fig. 2].

For the colonizers, maps were crucial in grasping the geography of their spheres of operation, desired and actual. Maps served a variety of important purposes. At the largest and most abstract scale, they sought to delineate as much of the continent as was known from empirical knowledge, and then often included much beyond, which ran the gamut from informed speculation to outright invention. Whatever their mixture of fact and fiction, such maps were powerful shapers of America in the European mind [Item 1.2]. They influenced monarchs and ministers as much as merchants and adventurers. It is generally accepted now that much of the information on such maps derived from Native Indian sources, particularly for regions beyond the personal reach of European explorers [Fig. 3]. Nevertheless, practically all of the maps produced in the context of European interest in North America were compiled, engraved, and printed by master cartographers far away in major European cities, often in the capitals of the kings and emperors who were their principal clients. These maps are in many ways artistic creations of the first order, and reflect the humanistic conjunction of art, artisanry, and developing geographical science in a period when wealth and ample time helped produce unique copies or limited editions of cutting-edge maps sought by an educated and powerful elite, many of whom had a direct interest in exploiting the possibilities of colonization. The exhibition made space for examples by Battista Agnese, Cornelis de Jode, and Vincenzo Coronelli [Items 1.2, 1.3, and 1.4].

While celebrated cases of Indian-made maps of varied territories and subject matter survive at different scales, much of the cartographic record of the Indians' estate has come down to us in the form of Euro-American recordings of Indian settlement patterns. Notwithstanding their outsiders' perspective and stylistic treatment, or perhaps even because of them (since they translate Native American geography by conventions easily understood in the Euro-American tradition), these images serve as suitable reminders of the prior colonization and long-established occupation of the continent's 'first peoples'. The exhibition includes an eighteenth century map of the territory held by the Cherokee Nation [Item 1.5], an 1853 watercolor drawing by Seth Eastman of the impressive hilltop Laguna Pueblo in New Mexico [Item 1.6], and John Wesley Powell's map of the *Linguistic Stocks of American Indians North of Mexico* [Item 1.7], a graphic reminder that the North American continent was not in any wise an empty, unpopulated land when Europeans arrived.

As European powers with colonial ambitions—such as Spain, France, England, and Russia—gained footholds on the continent, their demands on maps grew. Maps were needed to keep score of territorial appropriations, both for military planning and civil administration; for propaganda in the competition for colonial possessions; and for diplomacy, when unequal investments in the colonies or lopsided success on the battlefield flipped possession from one proprietor to another. These cartographic requirements can be symbolized at two scales, that of territory controlled by each colonial power, and that of a representative political or military stronghold symbolizing control over the territory.

Since Spain was first in the field and developed a far-flung empire in the New World distinguished by relatively centralized administration, it needed maps that documented this intensive investment. The 1729 Francisco Alvarez Barreiro map of the provinces of Coaguila and Nuevo Leon is the first to give an overview of this northern segment of New Spain covering much of the present-day southwestern United States [Item 1.9], and it is counterbalanced with images of the presidio town of San Antonio de Bexar [Items 1.8 and 1.10].

French and British competition was geographically so close that French maps, such as that by Guillaume Delisle [Item 1.12], went out of their way to belittle the British possessions, while British maps returned the insult, with interest [see the John Mitchell map, Item 1.15]. A rare manuscript sketch map by George Washington in his own hand of the territory through which he passed in 1753–54 (on behalf of the British to confront the French about their proliferating forts in the Ohio country) lends immediacy to this colonial rivalry [Item 1.16]. Washington's trip was the seed that spawned the French and Indian War, which then escalated into a conflict on European soil.

Russia's late entry into North America remained largely coastal, and at the cost of extraordinarily distended supply lines. New Archangel (later Sitka), Russia's colonial capital in what is now the southeast panhandle of Alaska [Fig. 4, Item 1.18], lay over 9,000 miles distant from St. Petersburg via the routes used, two thirds of that overland across Siberia. Nevertheless, Russian interests along the Pacific Northwest coast by the mid-nineteenth century are well illustrated in an atlas plate from 1848 [Item 1.19].

The site that would become Chicago emerged fitfully during the colonial period as a spot on the map, shown as a very short Chicago River [see pp. 50–51], later

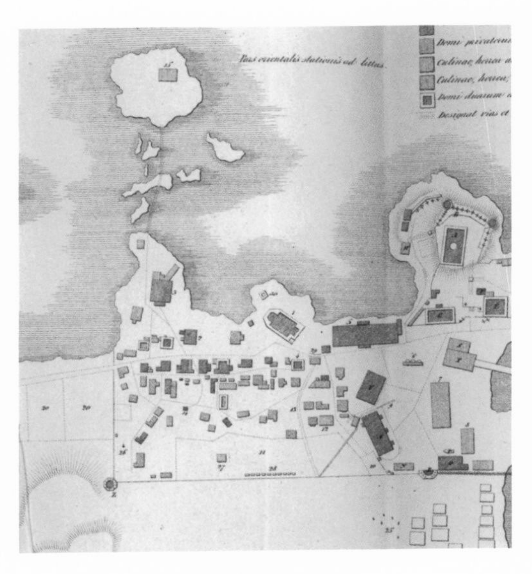

FIG. 4: Detail of Mashin 1842; see Item 1.18.

FIG. 5: Detail of Charles Preuss, *Topographical Map of the Road from Missouri to Oregon*. Baltimore: E. Weber & Co., 1846 (sheet 4 of 7). Newberry: vault map Graff 3360.

as a portage, and later still as a settlement with a fort. It was in this era that Chicago could last be reasonably described as lying in the Far West. More significantly, its representation as a transfer point across a critical continental watershed between the Great Lakes and the Mississippi Basin foreshadowed its future as a great industrial metropolis.

Mapping to build a nation

Once Anglo-Americans threw off the British yoke and formed a new confederation of independent states, the mapping of America took on a whole new complexion. Maps played a part in defining the boundaries of the new states, in charting the settled portions of the country, and in the encouragement of commerce. Private individuals carried out much of this activity, sometimes with support from advance subscriptions, occasionally with minimal public funds. New kinds of maps appeared everywhere, but the most dramatic developments involved the American West. By the early nineteenth century the concept of Manifest Destiny had taken root in American political discourse, although the phrase itself would not be coined until 1845. Jefferson's Louisiana Purchase had given impetus to the basic idea, and whether for reasons of propaganda or pragmatism maps immediately assumed a central role in framing newly-acquired knowledge about the West. The enormity of the new possessions, their relation to a fabled 'northwest passage' by water to the Pacific Ocean, and the prospect of untold riches to be gained from rocks, trees, and soil propelled maps into the limelight as legible and effective carriers of such information in a form that seemed almost glibly to suggest access. If it could be found on a map, it seemed virtually reachable.

The federal government underwrote much of this cartographic activity. Many of the maps were made on government time and printed by Congress for public distribution. In certain ways it could not have been otherwise. It was an acceptable use of the U.S. Treasury to pay for expeditions to record the West's geography, for development depended on understanding its structure and pattern. Where were mountains, and where were the passes through them that opened up new realms to settle and exploit? Where was water, and what collaterals did it possess? Such discoveries also paved the way to remove native peoples from their homelands, opening up great swathes of additional land for American settlers to colonize and transform. At first Western spaces were so vast that some reservations to which Indian tribes were herded and confined were quite substantial, but as colonization pushed westwards, they shrunk. Tribes were then relocated to lands the whites did not want, while reservations in more desirable locations were opened up to partition and sale by the Dawes Act.

The federal government was not the only source of maps that facilitated nation-building, but it was by far the most prolific and varied producer and sponsor. The earliest and most fundamental mapping activity was general exploration, to determine the character of terrain and identify corridors for passage and areas of land with mineral or agricultural potential. The most storied case was the Lewis and Clark expedition to the Pacific Northwest in 1804–06. The maps that resulted produced new and accurate information in rough proportion to distance from the route followed, but the improvement in general understanding of the Trans-Rocky Mountain West at that early point was substantial, as reflected in William Clark's 1810 manuscript compilation arising from his notes and reflections [Item 2.1].

There followed a series of landmark expeditions inspired by this success, many of them by army officers, such as Zebulon M. Pike, Stephen H. Long, Benjamin L. E. Bonneville, John C. Frémont, G. K. Warren, and George M. Wheeler. One of the most remarkable cartographic treasures to emerge from this process is the Frémont-Gibbs-Smith map, which comprises the 1844 edition of Frémont's map of the West and on which were later added copious pencil physiographic features, exploring routes, and other information derived from the long-lost manuscript maps of the great Mountain man, Jedediah Smith, made before his death in 1831 [Item 2.2]. Smith's knowledge would not be confirmed by other mapping syntheses until the 1850s.

Closely related to explorers' maps were those concerning military operations. Army field maps synthesized regional information at a large scale, with fully-developed topographic detail over very large areas. A rare manuscript case of this is the 1866 Grenville Mellen Dodge map of the Military District of Kansas and the Territories, designed to be carried on horseback [Item 2.8]. Military operations, of course, required follow-up with reports, a subset of which concern campaigns against specific Indian communities and detail the Army's role in their displacement and dispossession [Items 2.5, 2.6, and 2.7].

Another kind of mapping under government auspices concerned key transportation corridors; the Rocky Mountains presented a special obstacle. There were surveys of the routes of emigrant trails, such as the Oregon Trail (Fig. 5) and the Mullen Road, and later the great railroad surveys of the 1850s and 1860s, pursuant to Congress's interest in supporting the construction of one or more transcontinental railroads. All of this easily fit under the rubric of vital infrastructure in building up the nation and, of course, pursuing Manifest Destiny. Warren's map of routes for a Pacific railroad, published in 1854, places the summary features of the various detailed corridor studies on a single map [Fig. 6].

Other government maps focused on resources, and represent a further trend towards specialized cartography answering to broad economic interests. The 1857 James Hall and J. Peter Lesley map of the geology of the West was the first attempt to provide a comprehensive overview of the vast region's geological make-up [Item 2.3]. It would be followed in later years by detailed examinations of smaller regions, such as a special study of the Grand Canyon, with spectacular renderings of the rock formations [Fig. 7]. This last undoubtedly fired the curiosity of Americans everywhere and helped stimulate Western scenic tourism. Later surveys, like the Clarence King Survey, mapped complex bonanza mine workings, such as those of the 1870 Virginia Mine on the Comstock Lode [Item 4.9]. What constituted assessments of resources continued to broaden, especially as aridity became a critical issue. By the 1870s, the John Wesley Powell Survey was mapping at the scale of whole states to record the limits of lands that could be considered suitable for irrigation and pasture use, as exemplified in the case of Utah [Item 2.9].

Parallel with the foregoing mapping activity, the federal government carried on an altogether different survey and mapping program, the systematic nature of which, at the scale on which it was carried out, had no precedent in world history (far outpacing the scattered rural centuriation of the Roman Empire). The Ordinances of 1785 and 1787 sorted western lands in the public domain into a rectangular grid of townships and sections to provide a uniform land division system that extended, with a few interruptions, across the continent. Surveyors were required to prepare detailed maps of each 36-square-mile township, creating a national map set that has since been referred to as America's Domesday Book. The cumulative effect of this program across the American West is dramatically summarized in George Wheeler's 1879 map showing the advance of the land survey [Item 2.10], documenting the commodification of Western lands and their conversion from undivided communal resource to negotiable individual property.

The last of the big federal mapping traditions encompasses careful topographic mapping on a comprehensive basis. It represents the ultimate exploration of the continent's geography—not only the character of the terrain but also the transformations wrought by people. It started with mapping of the nation's coastlines by what became known

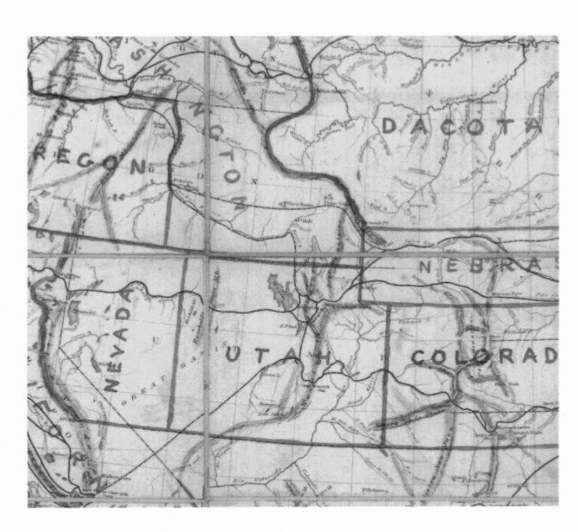

FIG. 6: Detail of Gouverneur Kemble Warren, *Map of Routes for a Pacific Railroad*. Washington, DC: Office of Pacific Railroad Surveys, 1857. Library of Congress.

FIG. 7: Detail of Holmes 1882; see Item 2.4.

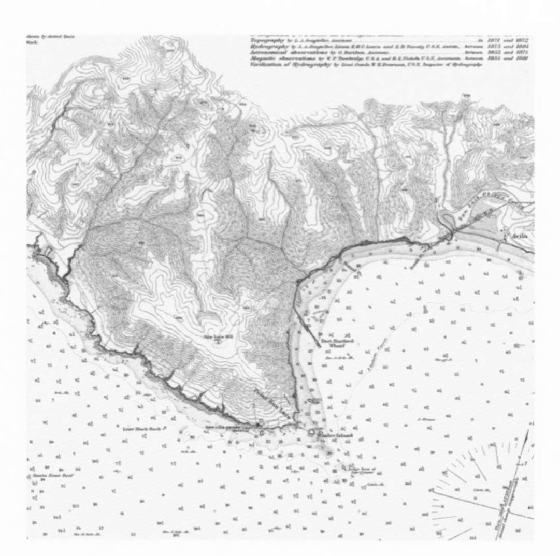

FIG. 8: Detail of USGS San Luis Obispo Bay 1889; see Item 2.12.

FIG. 9: USGS Chicago Quadrangle 1902; see Item 2.16.

as the U. S. Coast and Geodetic Survey, because of the risks to commerce of shipwrecks on uncharted shores. Some of the finest maps ever produced by the federal government can be found among these coastal charts [Fig. 8]. It would not be long before the demand for similar cartographic portrayal of inland areas gained strength, resulting in the creation of the U.S. Geological Survey. Starting in the 1880s under J. W. Powell's leadership, the *Topographic Map of the United States* required much of the twentieth century to complete basic coverage of the entire nation in the form of a giant mosaic of local sheets at a standard scale.

The United States was barely forty years old when Chicago changed almost overnight from métis village slumbering in the sunset of the Midwestern fur trade to American commercial upstart serving as outpost to the Eastern Seaboard mercantile complex. Chicago's beginnings are so firmly connected in the popular mind to its early commercial success and unrestrained private land speculation that the town's initial stage as a creature of government planning and economic largesse is often overlooked. As a nascent modern city it was platted by the State Canal Commissioners in 1830, placed shrewdly astride the confluence of the north and south branches of the Chicago River not far from Lake Michigan [Item 2.13]. While privately produced maps in New York and Philadelphia appeared to feed the real estate frenzy at this potent location, urgent matters such as harbor improvements required careful mapping by a federal government that was needed to implement them [Item 2.15]. As the city grew, local map publishers multiplied to cater to the demand for street maps, railroad maps, and other utilitarian products. Nevertheless, the federal government continued to play a part in the mapping of an emerging metropolis. It did so in ways that produced maps the commercial sector would never have attempted, such as the topographic sheet for the city in 1902, which coincidentally captured Chicago's astonishing physical size and complexity [Fig. 9]. Among other characteristics, such as scale and nationally systematic design, the map's depiction of building density reveals differences across city space and the interdigitation of industry and residential districts [Item 2.16] that private maps would scarcely have shown. With a history of government maps like these, Chicago, one could say, became official.

Maps for enlightenment

If government's primary concern for its citizens is their security and well-being, the maps most appropriate to those goals will provide fundamental information about territorial contents and promote economic development in the most general terms. Interpreting the findings of government geographical expeditions and surveys for the broad public was often left to the private sector, as were attempts to focus on the exploitation of particular resources. Private mapping, therefore, divided between that which had an educational purpose and that which directly aided business and the pursuit of profit. Maps that were made to enlighten their readers, while extremely diverse, tended to fall into one of three types along a continuum between impartial presentation and ideological argument.

The young Republic needed maps that informed its citizenry about the nation's regional geography, made sense of national expansion, and presented this information in simplified form for the rising generation—young children. Encyclopedic geographies and systematic atlases organized this knowledge, usually at the state level. The addition of Western territories to these atlases and geographies, as new states were admitted to the Union, gave the latest editions perpetual appeal. Emma Hart Willard, an early champion of teaching American history with maps (as well as of women's education) published original thematic maps that told a story of the development of civilization on the American continent. Her map of the wanderings of "aboriginal tribes" intended to show how shifting and impermanent their hold on territory had been before white colonization [Item 3.1]. Another map sought to present the new United States in 1789 as a set of modern states with fixed boundaries whose political development could proceed accordingly [Item 3.2]. In his 1835 map of the stages of civilization attained across the face of the globe [Item 3.3], William Woodbridge may have felt he was simply demonstrating a self-evident truth, but it is hard with today's eyes to miss his historical and ethnocentric prejudices [Figs. 10 and 11].

The theme of human development in specified geographical domains is continued in the textbooks and atlases of S. Augustus Mitchell, who became one of the leading publishers of such materials in the middle of the nineteenth century. Based in Philadelphia, which had a long history of printing and publishing and which was well located to reach the majority of Americans—a population still heavily concentrated on the Eastern Seaboard—Mitchell kept up with the rapid advance of geographical knowledge about the United States and routinely updated his coverage of Western regions [Items 3.4 and 3.5]. By the Civil War, Philadelphia publishers had stiff competition from others in New York, and by the 1880s from those in Chicago. Rufus Blanchard, a New England transplant to Chicago, issued educational maps for schools and home use, and his son-in-law, George F. Cram, would harness cerography, the cheap new printing technology based on wax engraving, to publish luxurious but affordable editions of atlases for home instruction [Item 3.6]. The title page of *Cram's Unrivaled Family Atlas of the World* focused unabashedly on all the key symbols of the new America of the West—a region that lay largely tributary to Chicago, now the most westerly of the great Eastern commercial and industrial cities.

Propaganda takes many forms, and the American West was the subject of intense propaganda during the nineteenth century because so many hoped to profit from it, directly and indirectly. A relatively disinterested example was the genre of the emigrant's guide, full of advice on what to expect, what to take, and how to get to the West. Maps in these primers offered some generalized geographical guidance as to best routes and landmarks along the way [Item 3.7]. While intended to be profitable for the author and publisher (often the same person), such guides filled a need for practical advice. Their very existence and proliferation only fed the popular image of the West as the Land of Promise, a promise that of course carried no guarantee from the author.

Cartography did more than just guide people to the West; it sometimes played a central role in formally enlarging it. John Disturnell's map of the United States in 1847, rather like the imperial maps of France and England a century before, sought to maximize the representation of American territory in a West still being fought over by the U.S. and Mexico. With the Treaty of Guadalupe Hidalgo, which temporarily settled the issue in 1848, the United States won more territory than it might otherwise have done when American negotiators used Disturnell's map and the Mexicans had none of their own with which to dispute it. Disturnell had the effrontery, cunning, or simple initiative, depending on one's point of view, to give the map a Spanish title [Item 3.8].

These visions of the West as the land of destiny were carried much further in other books and maps. A remarkable instance is provided by the Western promoter, William Gilpin, who designed a map to show that the American future lay emphatically in the West, that the gravitational pull of its natural resources would result in a population distribution so geographically equalized that the continent's greatest city would naturally develop somewhere in the vicinity of central Kansas, and that wealth and population would radiate from there in a series of boldly-marked concentric circles [Item 3.9].

But propaganda can push in various directions. Although maps generally portrayed

FIG. 10: Detail of Woodbridge 1835; see Item 3.3.

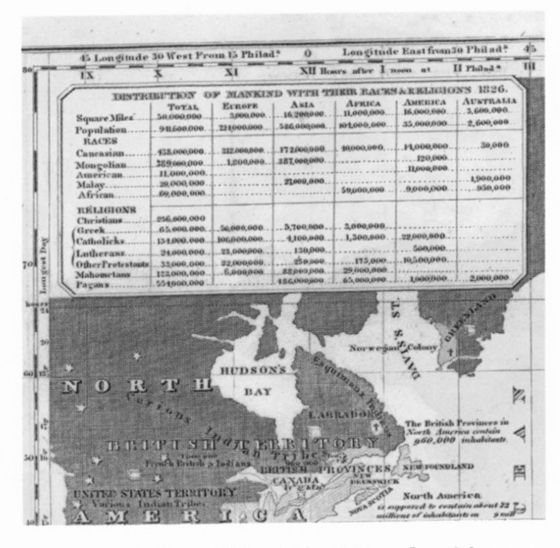

FIG. 11: Detail of legend, Woodbridge 1835; see Item 3.3.

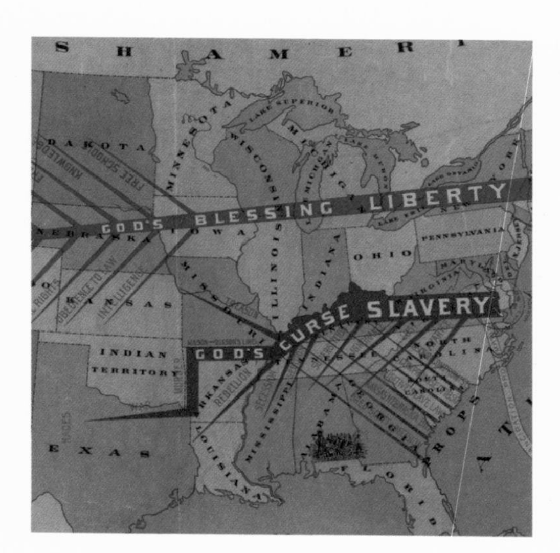

FIG. 12: Detail of Smith 1888; see Item 3.11.

FIG. 13: Detail of Thompson & Everts 1871; see Item 3.14.

the West as a region of seemingly limitless bounty, once in a while that theme could be harnessed to the cause of social justice. In a tirade that reflected the increasing anxiety felt in some quarters over the rise of America's first great corporations—the railroad combines—and the power they wielded, an 1874 map characterized the collective distribution of the extensive Western railroad land grants as a grand squandering of the nation's patrimony [Item 3.10]. Even more socially dogmatic was John F. Smith's *Historical Geography*, published in Chicago in 1888, which shows "God's curse of slavery," depicted as a jagged and misshapen tree whose branches list all its attendant evils, laid across and stunting the westward expansion of the South, contrasted with "God's Blessing of Liberty," in the form of a flourishing tree of virtues extending from sea to shining sea [Fig. 12].

Somewhere in between the educative and the propagandistic extremes of the cartographic continuum of maps produced for enlightenment were maps that chronicled the physical progress of Western settlements. These maps varied widely in scale, content, and detail, but they had a common theme: to show how American colonization was creating new places, filling them with the accoutrements of modern civilization, and providing individuals and families with the visual evidence of their of hard work and its effective transformation of the landscape. The theme is particularly well developed in maps deliberately made as historical testaments (of which all examples in this exhibition were made in Chicago).

In 1844 Juliet Kinzie published a partly autobiographical account of Chicago's early days, and included a map of the settlement as it had been in 1812 in order to show, by implication, how far it had come in just 32 years [Item 3.12]. Elsewhere, the interest in charting the growth and elaboration of towns is beautifully illustrated by Hogane and Lambach's impressive 1857 map of Davenport, Iowa, on the west bank of the Mississippi River, complete with "suburbs" [Item 3.13]. The lithographic views of prominent buildings and the generally decorative treatment of the map, not to mention the precision given to delineating the pattern of streets, urban lots, and full complement of buildings, all reveal the map's more-than-utilitarian purpose. It is a celebration of what the community had achieved in creating a thriving and physically complex settlement in its short history.

Documenting the accomplishments of a pioneer generation of settlers in a rapidly maturing new West was carried over to the rural realm by the western-style subscription county atlas, a cartographic art form particularly popular across the Midwestern and Plains states and in northern California during the 1870s and 1880s—wherever diversified farming and locally-based agricultural processing created a sufficient pool of potential subscribers. These atlases included detailed maps of every township within a county, showing individual land ownership parcels and acreage, within the matrix of local drainage, vegetation cover, roads, schools, churches, and other evidence of cultural development. They also contained often numerous and elaborate perspective sketches of individual farm properties, complete with crops and livestock in the fields, orchards, homes and farm outbuildings—in short, everything that signaled how successful the owner had been in wresting nature from the wilderness and creating a "middle landscape" of fruitful land use and material well-being. The DeKalb County, Illinois, atlas of 1871, by Thompson & Everts [Fig. 13], shows not only the farm spread of Joseph Farwell Glidden (on the map of DeKalb Township), but also a lithograph of his comfortable home [Item 3.14]. Glidden would shortly go on to further success as the inventor of the modern barbed wire design that revolutionized the American West.

This mapping genre was extended by Chicago publisher Alfred T. Andreas to books that attempted statewide coverage. In his celebrated *Historical Atlas of Minnesota* (1874), Andreas went all out to convey the impression of a frontier state rising fast to join its more developed sister states to the east by including a chromolithographic panorama view of the state capital, St. Paul, drawn by his top sketch artists, George H. Ellsbury and Vernon Green [Item 3.16]. While a few atlas subscribers saw in all this a form of advertising, the genre had far less to do with any direct aid to their businesses than with a belief in American progress, pride in making history, and personal achievement in a socially fluid and westering American society. All this subscription atlas publishing activity helped justify the construction and decades-long success of Chicago's Lakeside Building, America's only multi-tenant specialized mega-structure devoted almost exclusively to the printing and publishing of maps and atlases [Fig. 14].

Mapping for business

For every map of the West commercially produced to fulfill an educative or ideological purpose, at least ten more were designed as direct, practical aids to business. If there was an ideological basis for these, it lay in the promotion of capitalism and the wider pursuit of personal gain. Eastern centers of mapmaking had churned out maps of western regions useful to businessmen for decades. But as the nineteenth century wore on, Chicago's centrality within the continent, and its rapid gain of both specialized and of cheap, mass-produced map media put it in the driver's seat with regard to business cartography. It is not that Chicago mapmakers garnered a monopoly of such production—and publishers in other western cities certainly managed to enter the field too—but that they excelled in catering to the traveling needs of the general public and the vast army of salesmen who crisscrossed the country via the giant wholesale mart that Chicago itself had become.

The railroad was central to integrating the West into the mindset and the economy of the nation. Given the huge distances involved in crossing the continent, the vital problem with railroads remained their origins as separate companies operating over short distances between specific pairs of cities. It took time to build long-distance corporate rail networks to reduce the number of transfers between different operations, and maps informed passengers and shippers that efficient route systems for long-distance journeys were developing. As transcontinental rail travel became feasible following the Golden Spike in 1869, the race was on to develop and market the major rail corridors across the Rocky Mountains to the Pacific. The Atchison, Topeka and Santa Fe Railroad used mergers and cooperative agreements with more westerly lines to create a Pacific route from Chicago to California, as shown in their 'system' map of 1906 [Item 4.2]. The Northern Pacific Railway did the same for routes to the Pacific Northwest, in which it had an interest, by employing Poole Brothers, a large Chicago company oriented to high volume business printing, to produce a map [Item 4.1] showing its Chicago and Great Lakes connections to the East.

If maps were basic to making sense of transportation in the American West, they were hardly less essential to the business of real estate. Since at least the Middle Ages the concept of property has been fundamental to all modern colonization efforts. In the American West as elsewhere, maps were fundamental in recording the subdivision of property for legal purposes and for its sale to others. An especially intriguing case is the manuscript map of Carver's Grant in the Upper Mississippi region, dated circa 1825 [Fig. 15]. Believed to have been prepared as evidence in a petition to the U.S. Congress, this map shows a large portion of northwest Wisconsin subdivided into a grid pattern of rectangular land parcels, some of which are shown as sold, based on a supposed personal land grant to Jonathan Carver, a British military officer, from two Dakota chiefs well before the territory became American. The map, prepared by the English descendants of Carver and some of his alleged purchasers, was part of an effort to obtain recognition of the grant by the United States so the petitioners

FIG. 14: Lakeside Building (1871) from Andreas 1874; see Item 3.15.

FIG. 15: Detail of Carver's Grant 1825; see Item 4.3.

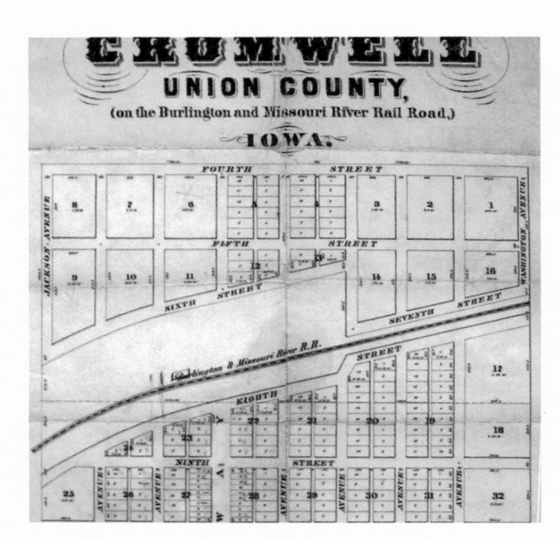

FIG. 16: *Map of Cromwell, Union County, Iowa.* Des Moines: Mills & Co. Lith., c. 1865. Newberry: sc CB&Q Map4F G4154. C7G46 1865 M5

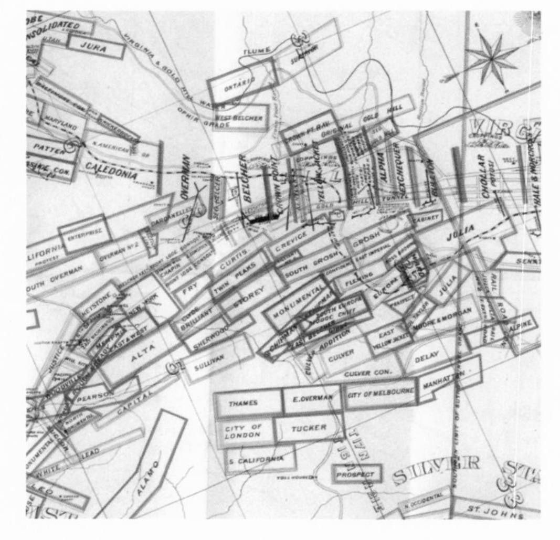

FIG. 17: Detail of Parkinson and Brown 1875; see Item 4.10.

could profit from further land sales within the grant's boundaries. As cunning camouflage, the map rather incongruously includes a patriotic portrait of George Washington. Congress investigated the claims made for the grant's historical legitimacy but found that no such grant could have been made under British colonial policy. What is interesting is how closely the grant's subdivision system mirrored what the U. S. General Land Office was developing as a standard for the American West in general, although it would be decades before federal surveyors would actually extend the land grid to this part of northern Wisconsin.

Equally intriguing is another map used in planning real estate strategy. An internal company document from the archives of the Chicago, Burlington, and Quincy Railroad, it specifies the additional lands outside its initial land grant corridor west of Omaha selected in lieu of the acreage denied it by the overlap of its grant with the prior one of the Union Pacific Railroad [Item 4.4]. The document consists of the surveyor general's official black and white printed map showing the progress of the U.S. Public Land surveys in Kansas and Nebraska as of 1865, on which the selected lands have been marked by a railroad official using several brightly colored inks. The survival of a unique manuscript map such as this one is quite rare. Vastly more numerous and widely known were maps published by the railroad companies for general distribution, inviting prospective settlers to purchase land from them rather than the U. S. government [Item 4.5]. Equally common, although more local and dispersed, were the numerous plats of new towns along the routes of anticipated railroads, aimed at spurring lot sales to settlers and speculators alike, such as the plan of Cromwell, in Union County, Iowa, astride the line of the Burlington and Missouri River Railroad [Fig. 16].

One last example of a map created to hawk real estate on a large scale is especially poignant. It concerns a 1900 map of the lands in Indian Territory assigned to the Creek and Seminole Nations but later opened up for possible purchase by whites through the allotment and private sale process [Item 4.6]. The map grades every land parcel into classes of desirability for agricultural use.

Western lands were valued not just for agriculture but for their potential mineral wealth. Maps appeared rapidly when mineral strikes hit the news and inspired stampedes. Hundreds of maps touted some aspect of the California Gold Rush, but few took such a comprehensive approach as the map published in 1849 by Ensigns and Thayer, showing not only the gold districts in California but the sea and land routes from the American East via Central and South America, as well as the overland routes straight west [Item 4.7]. Railroad companies were quick to publicize their routes whenever and wherever they led to the goldfields, and Chicago mapping firms soon participated in this development [Item 4.11].

Once there, other kinds of maps helped attract participation, record claims, and generally memorialize the mining environment. John Pratt and Bela Buell published a detailed map of the mountainous district around Central City, Colorado, in 1862, complete with dramatic hachures depicting the steep topography and the mines, roads, and other cultural features of the vicinity, lithographs of the mining town and many of its buildings, and an inset map of the state for context [Item 4.8]. The civil engineer T. D. Parkinson and his publishing partner G. T. Brown produced a map of the Comstock Lode in Nevada, delineating all the mining claims staked out on the ground [Fig. 17]. Not many years later, an extraordinarily detailed and colorful map of the subterranean workings along the Comstock Lode was published under the auspices of the Clarence King geological survey conducted along the Fortieth Parallel [Item 4.9]. While this was technically a government map, it recorded the results of the extensive commercial operations along the lode and was as much a salute to private enterprise as any privately produced map could have been.

One of the marvels of the American West was the ways it surpassed its economic projections. Everyone could imagine obtaining land there for farming, ranching, mining, and logging, ways of exploiting the West that drew sustained interest and investment. Anyone with knowledge of the West knew, too, that it was a palimpsest of natural wonders that appeared in sometimes fantastic and grandiose scenery. But it took government action and entrepreneurship both to commodify landscape for psychic consumption. Railroad companies found the key through their own urgent need to stimulate traffic any way they could. The natural wonders of Yellowstone were brought within the protective management of the nation's first national park in 1872. Soon the railroads began developing package tours to the Rocky Mountains, thereby creating a tourist industry that needed maps both to orient visitors to scenic attractions and to make their visits practical. Hence, a new kind of map emerged, sometimes wrapped in the imaginative clothing of a child's adventure story [Item 4.12]. Over time, as interest increased and venues multiplied, the presentation of scenic information became more matter-of-fact and sophisticated [Item 4.13].

No one city has ever dominated the cartographic images of so large a region, but Chicago's evolving fortunes allow us to chronicle and assess how one place can shape the image of so consequential a world region as the American West—a region that Chicagoans worked to make tributary to their city. Chicago mapmakers played their part in shaping the region's national image by participating in all the major developments in business cartography that affected the West, and as the nineteenth century gave way to the twentieth, a number of their innovations became national standards for decades to come.

Inevitably in an entrepreneurial environment, some experiments failed to catch on. A particularly interesting early case is provided by fire insurance mapping. Commerce expands when risk is reduced. As property insurance became central to business operations, and insurance companies arose whose customers might have buildings anywhere, detailed maps of the built environment in major cities became a distinctive form of business mapping. The earliest known fire insurance atlas of an American city was the simplistic 1861 Pinney atlas of central Boston. Just four years later, with no other precedent to stimulate him, a Chicago insurance agent was also in the game. B. W. Phillips employed Frederic Cook to create a singularly detailed atlas of central Chicago, including "autographic drawings" of numerous important buildings. This technique blended simple block plans with elevational building profiles laid flat across the structure footprints, showing the number of stories and cornice designs [Fig. 18]. The style is unique, and would have proven unworkable in later years as skyscrapers grew taller and taller, but the effect captures in dramatic and extraordinary detail the changing character of Chicago's rapidly expanding commercial core six years before the Great Chicago Fire of 1871. In the post-fire era Charles Rascher and others made fire insurance atlases of Chicago, but eventually these local efforts withered as the Sanborn Company came to dominate insurance mapping nationwide. Cook's atlas, nevertheless, is a testament to the experimental and energetic mapping culture taking shape as Chicago—then a precocious Western city—grew by leaps and bounds.

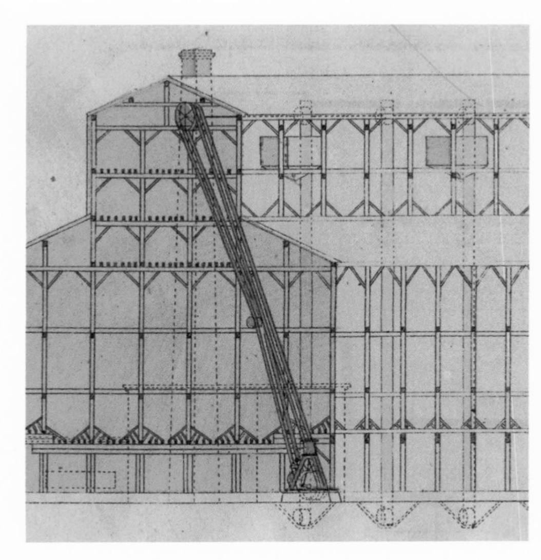

FIG. 18: Detail of Cook 1865; see Item 4.14.

Other Chicago innovations had far more lasting impact on western and national mapping. No firm is more closely associated with this phenomenon than Rand McNally and Company. Begun in 1856 primarily as a railroad ticket printing operation for the lines that radiated from Chicago, the firm quite logically moved into the production of railroad maps showing all the stations their routes connected. From this it was a short step to developing state maps and, by 1876, an atlas portraying every location

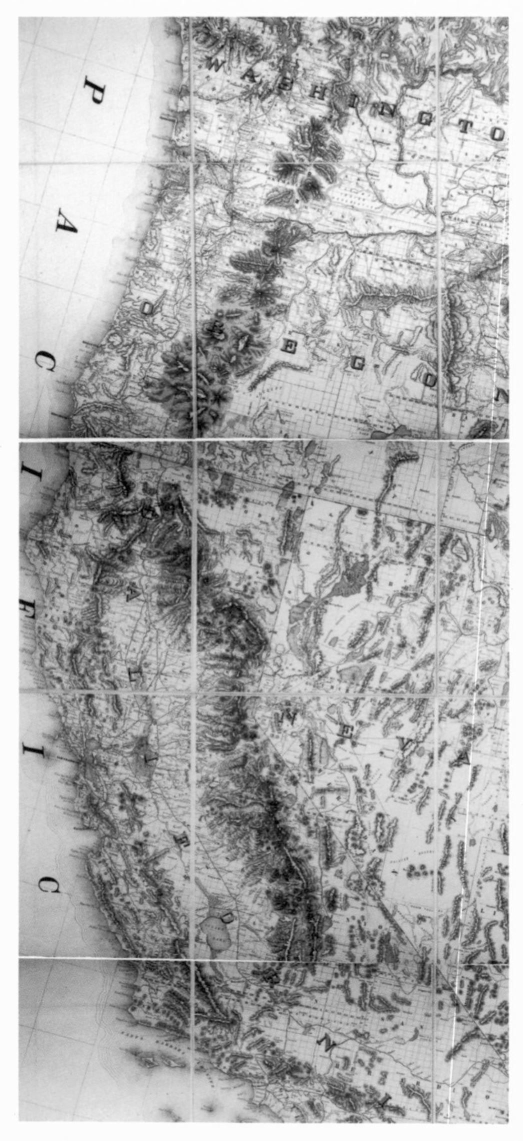

FIG. 19: Detail of Rand McNally 1887; see Item 4.17.

that travelers and business representatives could reach by rail. Titled a "business atlas," this cartographic formula met a large and expanding need, particularly as settlement intensified in the American West [Items 4.15 and 4.16]. In an era when traveling salesmen roamed vast territories in search of sales, an atlas that provided vital geography and population data helped plan their attack. Printed by the cerographic method, which made it cheap and easy to update maps to reflect new places, especially in the rapidly developing West, the firm of Rand McNally, created a perennial cartographic product. It provided a firm financial basis for a growing array of maps for all kinds of business customers that established the company as the leading mapping firm in the nation by the middle of the twentieth century. They were not elegant—certainly a far cry from the exquisite early copperplates printed by the U. S. Coast Survey or the chromolithographic maps issued by some of the great federal Western surveys—but these functional and inexpensive maps produced in Chicago for businessmen everywhere played their own vital part in underwriting the expansion of the West and its integration in the national economy.

Nowhere was this on more glorious display than in the "monster map" that Rand McNally published in 1887. Perhaps the largest commercially produced map ever printed in the United States, *Rand, McNally & Company's Standard Map of the United States of America, with Portions of the Dominion of Canada and the Republic of Mexico* stretches nine feet tall by fourteen feet wide [Fig. 19, Item 4.17]. Produced as a marquee map to hang on the walls of boardrooms and the offices of company presidents seeking to dramatize their national and international commercial ambitions, or perhaps in overseas embassies, the map brings the enormity of the United States as a geographical reality sharply into focus. The map scale, at slightly more than one to one million (1:1,140,480), permits every single post office and way station on the continent to be shown in relation to natural features such as topography and drainage; it shows fundamental cultural features such as the land survey system, political divisions, and the entirety of the all-important railroad network. Stand close enough to the map to take in this amount of detail and one cannot but be awed by the power of the cartography. It invites one to locate a familiar place and then, in one sweep of the eye, to relate that place to the gigantic, complex continent in which it is situated. The map fairly screams the message of Manifest Destiny realized, with a confidence that seemed almost to predict the country's rise to world power in the century that followed. From the tentative maps of the American West of just 75 years earlier, full of holes reflecting their dearth of geographical knowledge of the region, to the Rand McNally monster map of 1887, with its cartographic evidence of the geographical integration of the West into the lifeblood of the American people, we have perhaps the ultimate symbol of a national dream that came true.

SECTION ONE: MAPS FOR EMPIRE

The earliest cartographic depictions of the American West appeared on maps of North America made by or for European imperial states during the early modern age of exploration. The establishment of American colonies was an extension of the competition among these powers for political and economic hegemony around the world between the sixteenth and nineteenth centuries. European governments commissioned sophisticated maps enabling them to comprehend the scope of North America and visualize their colonial ambitions. As European powers vied for control of the continent, maps helped to establish boundaries, publicize territorial claims, and recruit settlers. Maps frequently acknowledged the presence of Native Americans, but not their rights to the land.

The examples in this section illustrate various national styles of both cartography and imperialism. They demonstrate how empires were created, practically and imaginatively, through maps. European declarations of ownership of the continent foreshadow the American proclamations of manifest destiny that spurred westward expansion in the nineteenth and twentieth centuries.

America in the European Mind

In the late fifteenth and early sixteenth centuries, the explorations of Christopher Columbus, Bartolomeu Dias, and Vasco da Gama introduced Europeans to an expanded world. Improved maps facilitated voyages, recorded their findings, and communicated this knowledge to the public. The earliest explorers saw America as a way station en route to the riches of Asia. In the seventeenth century, Europeans began to acknowledge the land's potential for settlement and exploitation. Maps introduced Europeans to the geography and economic promise of the continent, stimulating desires for conquest and wealth. They also stoked European imaginations, prompting literary and pictorial explorations of exotic places.

1.1

A Renaissance Sailor's View of the Americas

CREATOR: Diogo Ribeiro

TITLE: *[Map of North America]*, 1529 (as reproduced in J[ohann] G[eorg] Kohl, *Die beiden ältesten General-Karten von Amerika, ausgeführt in den Jahren 1527 und 1529 auf Befehl Kaiser Karl's V in Besitz der Grossherzoglichen Bibliothek zu Weimar. Erläutert von J.G. Kohl.* [The two oldest general maps of America. Executed in the years 1527 and 1429 by command of Emperor Charles V. From the grand-ducal library in Weimar.]) Weimar: Geographisches Institut, 1860)

SIZE: 36 x 27.25 inches

LOCATION: Map Collection, Regenstein Library, University of Chicago [not on display]

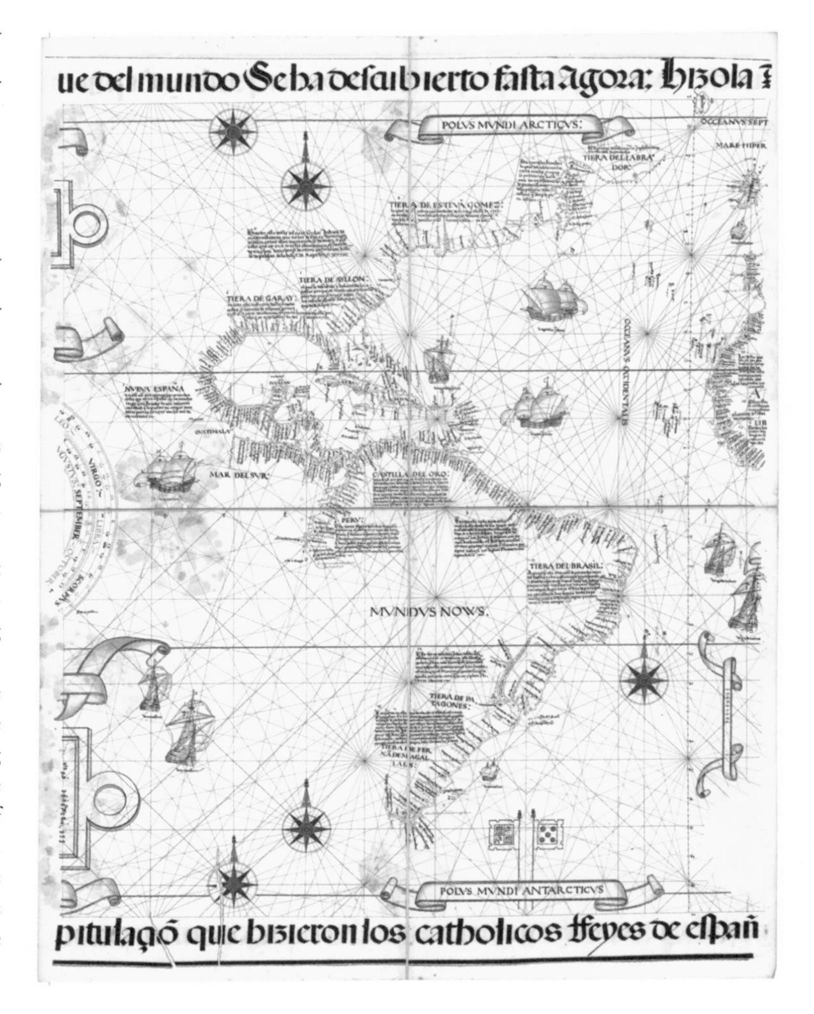

Diogo Ribeiro's 1529 map of the world is regarded as one the finest maps of its era. The Portuguese-born adventurer developed his sea legs at a young age, serving as a pilot on several voyages to India. He began serving the Spanish crown in 1518 or 1519, and in 1526 King Charles V appointed him Pilot Major, with instructions to interview returning navigators and incorporate their latest geographical findings on the official *padrón general*, a master world map. Ribeiro depicted the world in the form of a portolan chart, a navigational map featuring a network of rhumb lines to indicate sailing directions.

By outlining the entire Atlantic seaboard in detail, the map illustrates the knowledge of the Americas gained since Columbus's 1492 voyage. For the northeast, Ribeiro relied on information collected by Esteban Gómez during his 1524 search for a western passage to China. For the southeastern coast, he drew on reports from Lucas Vasquez de Ayllon's failed Carolina settlement of 1526. To compensate for the lack of information about the continental interior, Ribeiro decorated the uncharted space with enlarged place names, compass roses, trees, and animals. Many subsequent cartographers, artists, and writers would take similarly imaginative approaches to depicting the land that became the American West.

Ribeiro revealed his professional breadth through the map's other decorations: sea-faring ships point to his experience as a navigator, while the astrolabe and quadrant in the lower corners and the scale of magnetic variation at the far left allude to his background as an instrument-maker and cosmographer. With these images, Ribeiro asserted the empirical basis of his worldview.

The sheet reproduced here is a chromolithograph of the 1529 manuscript, published in Johann Georg Kohl's 1860 *Die beiden ältesten general-karten von Amerika*. Kohl, a German travel writer and geographer, was an important scholar of the early exploration of North America. He and his assistants prepared hand-drawn facsimiles of early maps of America held in European libraries, many of which are now housed in the Library of Congress. Members of his family immigrated to the United States and settled in Chicago, where his nephew, Charles E. Kohl, met success as a theater impresario.

References: Cumming, Skelton, and Quinn 1972; Glover 1911; Mollat du Jourdin 1984; Nebenzahl 1990; Stevenson 1909; Vigneres 1962; Winsor 1904; Wolter 1981

1.2

Territorial Riches, Cartographic Luxury

CREATOR: Battista Agnese

TITLE: [*Map of North America*] in Battista Agnese [*Manuscript Portolan Atlas of the World*], c. 1550. Pen and ink with hand coloring, on vellum

SIZE: 11 x 15.25 inches

LOCATION: Newberry vault Ayer MS map 12

CREDIT: Gift of Edward E. Ayer

Battista Agnese (fl. 1536–1564) produced some of the most beautiful portolan atlases ever made, matching the artistic achievement of the most sophisticated illuminated manuscripts of his day. Delicately painted in luminescent pigment on thick vellum, Agnese's luxurious materials evoked the riches explorers hoped to find in the places he depicted. He made these sumptuous atlases not for seafarers, but for wealthy collectors. Kings, popes, diplomats and nobles all owned portolan atlases, prizing them as status symbols as well as reference tools.

Agnese was one of the most prolific cartographers of the sixteenth century. He identified himself on his maps as Genoese, but his career flowered in Venice between 1536 and 1564. Agnese's maps, like most of the era's Italian portolan atlases, did not convey new information. Lacking first-hand knowledge of places he depicted, he synthesized the maps produced by others, relying heavily on Portuguese and Spanish models.

His map of the southwestern section of North America is one of the oldest surviving representations of what would become the Western United States. Agnese incorporated information from recent Spanish explorations, including place-names recorded by Francisco Coronado during his expeditions of 1540 and 1542. Coronado was drawn north by descriptions of the legendary seven cities of Cíbola and the province of Quivira, whose wealth was believed to rival that of the Aztec capitals. Agnese's map is also notable as one of the earliest reasonably accurate representations of the California coast, based on Francisco de Ulloa's explorations in 1539 and 1540. Subsequent European cartographers, including Vincenzo Coronelli (see Item 1.4), would show California as an island.

References: Buisseret 2007; Campbell 1987; Cohen 2002; Wagner 1931; Wallis 1997; Woodward 1997

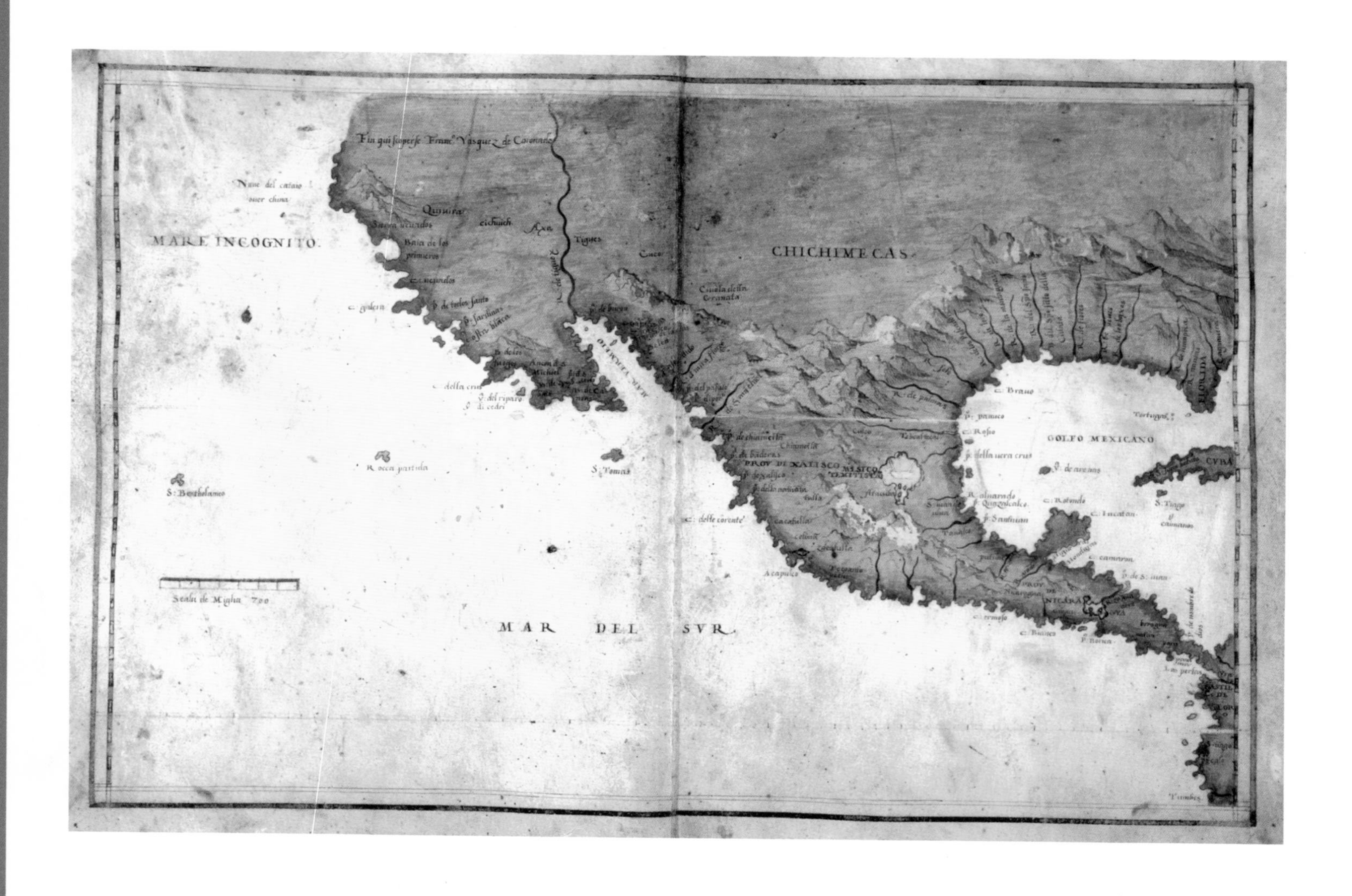

1.3

America in the Early Modern World Atlas

CREATOR: Cornelis de Jode

TITLE: *Quivirae regnum cum aliis versus boream* (*Realm of Quivirae with Others Toward the North*) and *Americae pars borealis, Florida, Baccalaos, Canada, Corterealis* (*The North Part of America, Florida, Baccalaos, Canada, Corterealis*), from Gerard de Jode and Cornelis de Jode, *Speculum Orbis Terrarum* (*Mirror of the World*) (Antwerp: Arnout Coninx, 1593)

SIZE: America: 16 x 21 inches; Quivira: 16 x 10.5 inches

LOCATION: Newberry vault oversize Ayer 135 J9 1593 (catalog reproduction); The Arthur Holzheimer Collection (exhibit)

CREDIT: Gift of Edward E. Ayer (Newberry copy)

Gerard de Jode
Hendrik Goltzius, *Portret van Gerard de Jode*, inv.: OP 7887
CREDIT: Plantin/Moretus Museum, Antwerp

In the later sixteenth century, a new consumer good reached the European market: the deluxe world atlas. Abraham Ortelius and other mapmakers in the Low Countries, then the leading center of cartography, pioneered the genre. Copies of Ortelius's elegantly bound, luxuriously printed *Theatrum Orbis Terrarum* (*Theater of the World*) attested to their owners' learning, taste, and worldly sophistication. Enterprising cartographers and publishers often sold the same maps individually as loose sheets before and after issuing the bound books. Their more modest editions made maps and geographical descriptions of places around the world available to wider audiences.

Gerard de Jode (1509–1591), a native of Nijmegen, the Netherlands, began his career as a cartographer, engraver, printer, publisher, and printseller in Antwerp around 1550. The sumptuous atlas he compiled, the *Speculum Orbis Terrarum*, appeared in 1573. After his death, his widow and his son Cornelis (1568–1600) continued his business. In 1593, Cornelis de Jode issued an updated edition of the atlas. He revised several maps and added some new ones, including the maps of North America and Quivira. The second edition sold better than the first, but neither sold as well as Ortelius' more comprehensive *Theatrum*. Both editions are rare today.

In contrast to Ribeiro's blank interior, de Jode filled North America with topographical features and place names. Although de Jode consulted the latest sources, including Theodor De Bry's two-volume *America* from 1590–91, much of the geography remained speculative. The broad body of water stretching across the north of the continent reflects Europeans' tenacious hope for the discovery of a westward route through America to China. By including a separate map of Quivira, de Jode perpetuated the fable of a much sought-after North American province that lured explorers for at least a century beginning in the 1530s. Quivira appears in various locations on early maps of the continent. Whereas Agnese recalled Coronado's search for the province in the area north of the Arkansas River (in what is now Kansas), de Jode shifted the site of Quivira to the Northwest coast. His depiction, which rather uncannily suggests present-day Alaska, furthered his view of the continent as a bridge to Asia.

References: Heidenreich and Dahl 1980; Krogt 1997–; Shirley 1983; Skelton 1965

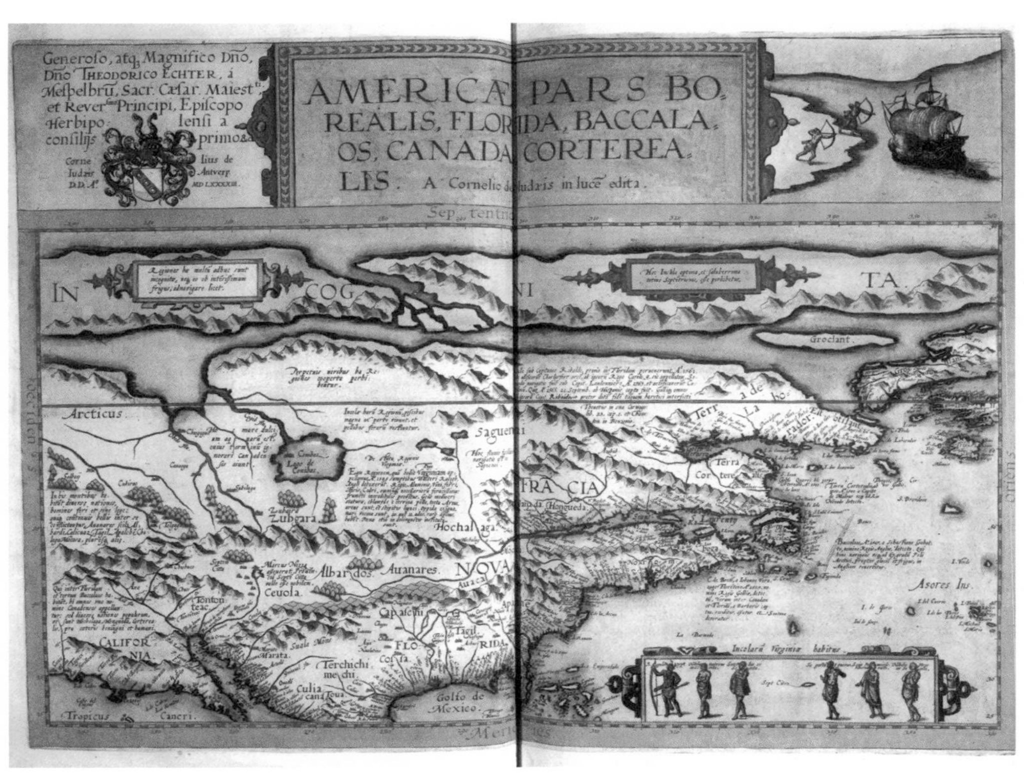

1.4

"Chekagou" Appears on the Map

CREATORS: Vincenzo Maria Coronelli (compiler), Dom Padovani (publisher), Anton Felice Marsili (dedicatee)

TITLE: *America Settentrionale* (*North America*), 1688, in Vincenzo Coronelli, *Atlante Veneto* (*Venetian Atlas*) (Venice: D. Padovani, 1690) vol.1, plates [14–15]

SIZE: 28 x 19.3 inches

LOCATION: Newberry vault oversize Ayer 135 C8 1690 vol. 1 (catalog reproduction); The MacLean collection (exhibit)

CREDIT: Gift of Edward E. Ayer (Newberry copy)

Vincenzo Coronelli

Atlante Veneto... (1691–1696), vol. 2, pt. 2 (maps C.44.f.6)

CREDIT: The British Library, London

Chicago ("Chekagou") first appeared as a place name on a printed map on a 1688 sheet, which Vincenzo Maria Coronelli (1650–1718) included in his monumental atlas, the *Atlante Veneto.* More broadly, the map titled *America Settentrionale* provided the most reliable depiction of the American West before the eighteenth century, representing the extent of European geographical understanding.

A Franciscan friar from Venice, Coronelli became cosmographer to the Republic of Venice in 1685. He relocated to Paris from 1681 to 1683 to produce a pair of monumental globes for King Louis XIV. His court position gave him access to the latest manuscript maps of North America sent back to France by explorers. Coronelli benefited from the charting of the Mississippi by Louis Jolliet and René-Robert Cavelier de La Salle, and the mapping of the lower Great Lakes by Jesuit missionaries. For the southwest, he relied on a small sketch by Diego Peñalosa, a former governor of New Mexico exiled in Paris.

Although much of its geography was accurate, *America Settentrionale* featured a few notable errors. Coronelli repeated La Salle's mistake in placing the mouth of the Mississippi 600 miles too far to the west. He located the Ohio River too far to the south, and misplaced an illustration of men killing alligators in the central Mississippi Valley. Most conspicuously, Coronelli depicted California as an island, citing in the cartouche several sixteenth-century Spanish explorations of the Pacific coast. His representation reversed the advance made by Agnese (see Item 1.2) and other sixteenth mapmakers, who attached California to the mainland. This regression dates back to the beginning of the seventeenth century, when several cartographers showed California as an island, fueling a geographic myth that persisted into the early eighteenth century.

References: Cumming 1974; Heidenreich 1980; Kaufman 1989

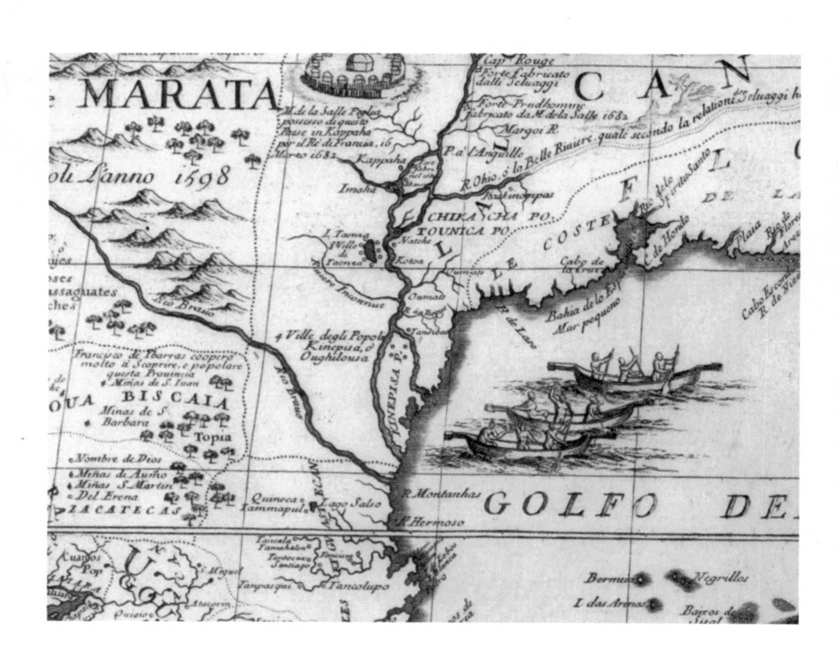

The Native American Landscape

Before Europeans arrived, Native Americans seem to have made maps for purposes ranging from planning travel to leaving messages, for religious worship, and for commemorating events. Drawn in dirt, painted on rock, or carved on blazed trees, most of these indigenous maps were ephemeral. Consequently, few maps made north of Mexico before European contact survive. Most of the Native American maps that do survive were made for Europeans. Explorers and settlers relied extensively on the natives' geographic knowledge, frequently asking them to make maps.

Although Indians did not use maps to demonstrate ownership, empire-minded Europeans adapted native maps to chart the lands they claimed. After the United States gained independence, the federal government mapped native homelands to further its goals of geographical expansion and Indian assimilation. By the end of the nineteenth century, officials no longer viewed Native Americans as foreign nations or military threats, and the government transferred Indian affairs from the War Department to the Department of the Interior.

1.5

A British View of the Cherokee Nation

CREATOR: Thomas Kitchin

TITLE: *A New Map of the Cherokee Nation,* in *London Magazine* vol. 29 (February 1760): 97

SIZE: 10.25 x 8.75 inches

LOCATION: Newberry A 51 .533

The *London Magazine* published Thomas Kitchin's *New Map of the Cherokee Nation* to illustrate a news report of the expedition of William Henry Lyttleton, the colonial governor of South Carolina, against the Cherokees. Mobilizing 800 militia and 300 regular troops, Lyttleton "intimidated the nation" and forced them to sign a treaty "conceived in terms most advantageous to the British interest." The article rationalizes the British domination of the Indians by noting the recent scalping of several colonists. The larger motive behind Lyttleton's campaign was the military importance of the Cherokees to Britain's colonial ambitions. The British were eager to secure Indian allegiance in the competition with the French for control of North America. Observing that the Cherokees "are the most numerous nation of Indians adjoining to the British colonies on the continent," the article asserts that they can muster "2,500 fighting men."

Thomas Kitchin (1719–1784) was a prolific London mapmaker and engraver. He engraved hundreds of maps for the *London Magazine* and other publications, and was appointed hydrographer to King George III by 1773. Kitchin never traveled to America. He based his *New Map of the Cherokee Nation* on an Indian-made drawing sent to London from the colonies. The map illustrates the size and strategic location of the Cherokee territory, extending from the western frontier of the British settlements in Virginia and South Carolina to the French settlements along the Mississippi. Although identifying the Cherokees as a nation imposes a European political category, the abundance of Indian names on the map attests to the Cherokees' well-established roots in the territory.

Lyttleton's defeat of the Cherokees was an early instance in a long history of subjugation by force. Beginning in the early nineteenth century, United States policy pushed Indians to adopt Euro-American ways. The Cherokees succeeded famously, becoming prosperous farmers, adopting a republican constitution, and declaring their independent nationhood in 1827. But few U.S. political leaders were willing to accept their sovereignty, and residents of Georgia in particular coveted the Indians' productive land. The discovery of gold on Cherokee land in 1829 intensified that desire. The following year, the Indian Removal Act signaled a shift to a new policy that ultimately forced the Cherokees to move to Indian Territory west of the Mississippi.

References: Anon. 1760; Hine and Faragher 2000; Prucha 1984; Prucha 1990; Worms 1993

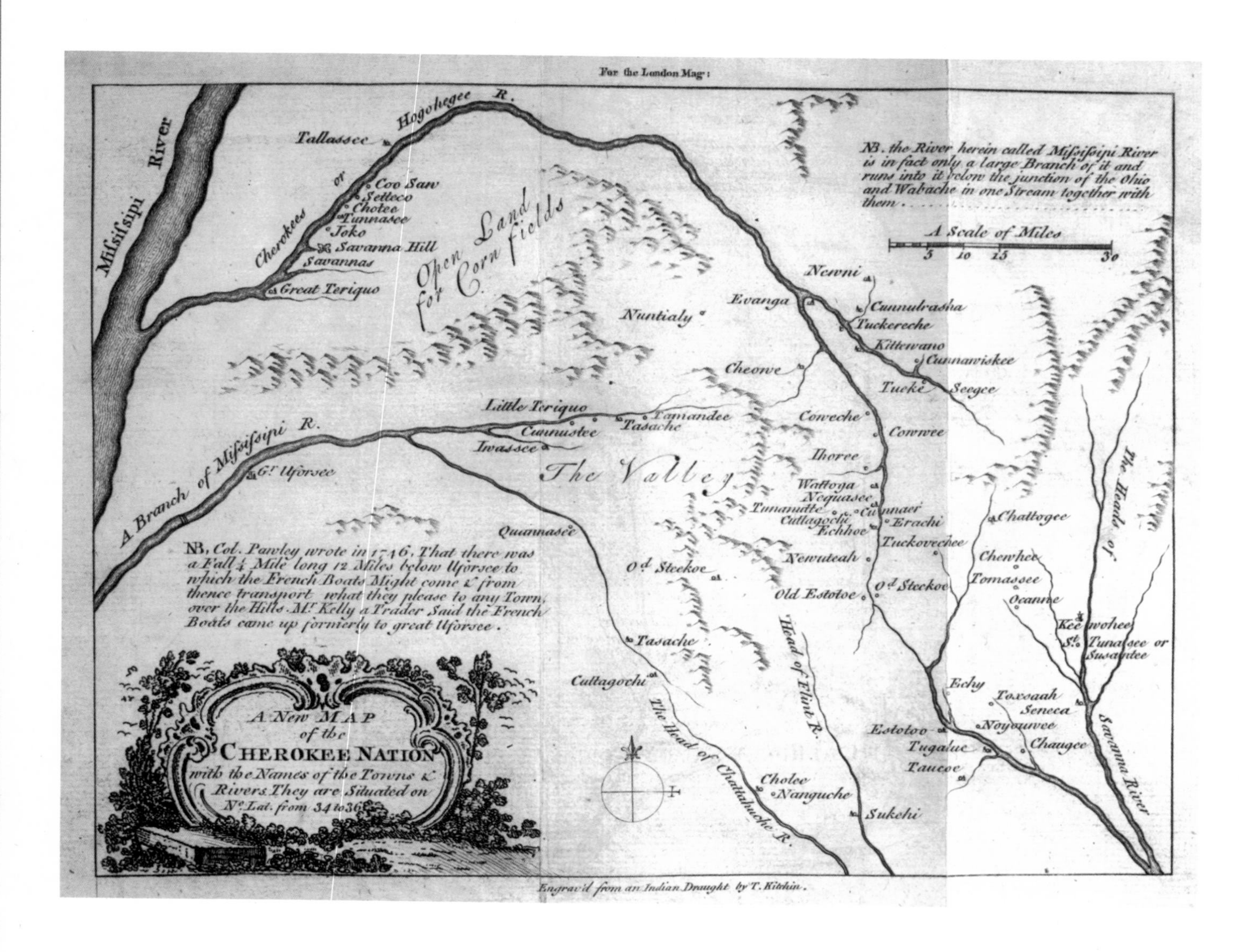

1.6

A Soldier-Artist Pictures Pueblo Culture

CREATOR: Seth Eastman

TITLE: *Pueblo of Laguna, New Mexico, Drawn by Capt. Eastman U.S. Army from a Sketch by R. H. Kern,* 1853. Watercolor on paper

SIZE: 12.6 x 15.7 inches

LOCATION: Newberry vault oversize Ayer Art Eastman

CREDIT: Gift of Edward E. Ayer

Seth Eastman
Seth Eastman in uniform of Brigadier General
CREDIT: National Anthropological Archives, Smithsonian Institution (OPPS NEG. 48185)

The land that is now central New Mexico was home to a succession of cultures as far back as 3000 B. C. The Spanish moved north from Mexico into the area in 1540, eventually prompting the Pueblo Indians to unite and repel them in 1680. When the Europeans returned in 1692, about a hundred Indians fled to the isolated Acoma Pueblo. In 1697 a smaller group migrated north to establish the Laguna Pueblo; the next year this group offered to make peace with the Spaniards. In the eighteenth and nineteenth centuries, Spanish ranchers encroached on Laguna land, while Apache and Navajo raided the pueblo for sheep and horses.

The Lagunas built in a style typical of the southwest, laying stone into wet clay to shape a cluster of contiguous rooms built into the upper slopes and over the top of the rocky hillside. Many of the historic structures remain today on their 420,000-acre reservation. After the discovery of uranium on the reservation in the 1950s, the Laguna became the wealthiest of the Pueblo groups in New Mexico.

A West Point graduate and career army officer, Seth Eastman (1808–1875) became a successful painter as well when his frontier postings brought him into close contact with Native Americans, who would become his primary artistic theme. In 1850, the Army detailed Eastman to the Bureau of Indian Affairs in Washington, D.C., to illustrate Henry Rowe Schoolcraft's *Historical and Statistical Information Respecting the History, Condition and Prospects of the Indian Tribes of the United States.* Published in six volumes between 1851 and 1857, the text resounded with the popular expansionist rhetoric of its day, maintaining that it was the Indians' duty to submit to the nation's Manifest Destiny.

Eastman spent five years drawing more than 300 sketches for Schoolcraft's history. The images attracted great critical praise when they appeared in the lavishly produced volumes. Eastman's view of the Laguna Pueblo looks north, picturing the complex from the south side of the San José River. He based the picture on a small wash drawing of the pueblo made by his friend Richard Kern, probably during the 1849 military expedition commanded by John Washington, territorial governor of New Mexico (see Item 2.5).

References: Hine 1982; McDermott 1961; Nabokov and Easton 1989; Ortiz 1979; Tyler 1987

1.7

The Distribution of Indian Languages Across the Continent

CREATOR: John Wesley Powell

TITLE: *Linguistic Stocks of American Indians North of Mexico,* in J[ohn] W[esley] Powell *Indian Linguistic Families of America, North of Mexico* (Washington, DC: Government Printing Office, 1891)

SIZE: 22.5 x 28 inches

LOCATION: Newberry Ayer 402 P8 1891

CREDIT: Gift of Edward E. Ayer

John Wesley Powell

Critic 41 (November 1902): 399.

CREDIT: Newberry A5 252

Like Kitchin's map and Eastman's watercolor, this map documented the Indians' historic presence in North America, illustrating the diversity and broad geographical reach of indigenous cultures. Identifying 58 linguistic families as they were dispersed before the arrival of European explorers, the map also epitomizes an important shift in prevailing attitudes toward Native Americans. As its author John Wesley Powell (1834–1902) put it in 1874: "There is now no great uninhabited and unknown region to which the Indian can be sent. He is among us, and we must either protect him or destroy him."

A Midwestern teacher and Civil War veteran, Powell gained expertise in the natural history and geography of the West by leading scientific explorations of the Colorado River and Rocky Mountains in the late 1860s and 1870s. In the field Powell developed an interest in Indians, who soon became the focus of his career. In 1879, he successfully lobbied Congress to establish the Bureau of Ethnology within the Smithsonian Institution and was named its first director.

Much as the Western landscape beckoned explorers, anthropology loomed as an intellectual frontier inviting systematic investigation. Powell saw the study of Indian languages as the heart of ethnology. He believed linguistics could provide insight into Native Americans' ideas and beliefs, which was pivotal to assimilation and to the shaping of a larger science of culture. At the Bureau, Powell worked with Henry Henshaw to synthesize data collected by a corps of field investigators. Their work culminated in the 1891 report, *Indian Linguistic Families of America, North of Mexico,* which included the *Linguistic Stocks of American Indians North of Mexico.* In 1893, the Smithsonian Bureau of Ethnology displayed the map as the centerpiece of its exhibit at the World's Columbian Exposition in Chicago.

References: Bartlett 1962; Darnell 1969; Darnell 1971; Hinsley 1981; Worster 2001

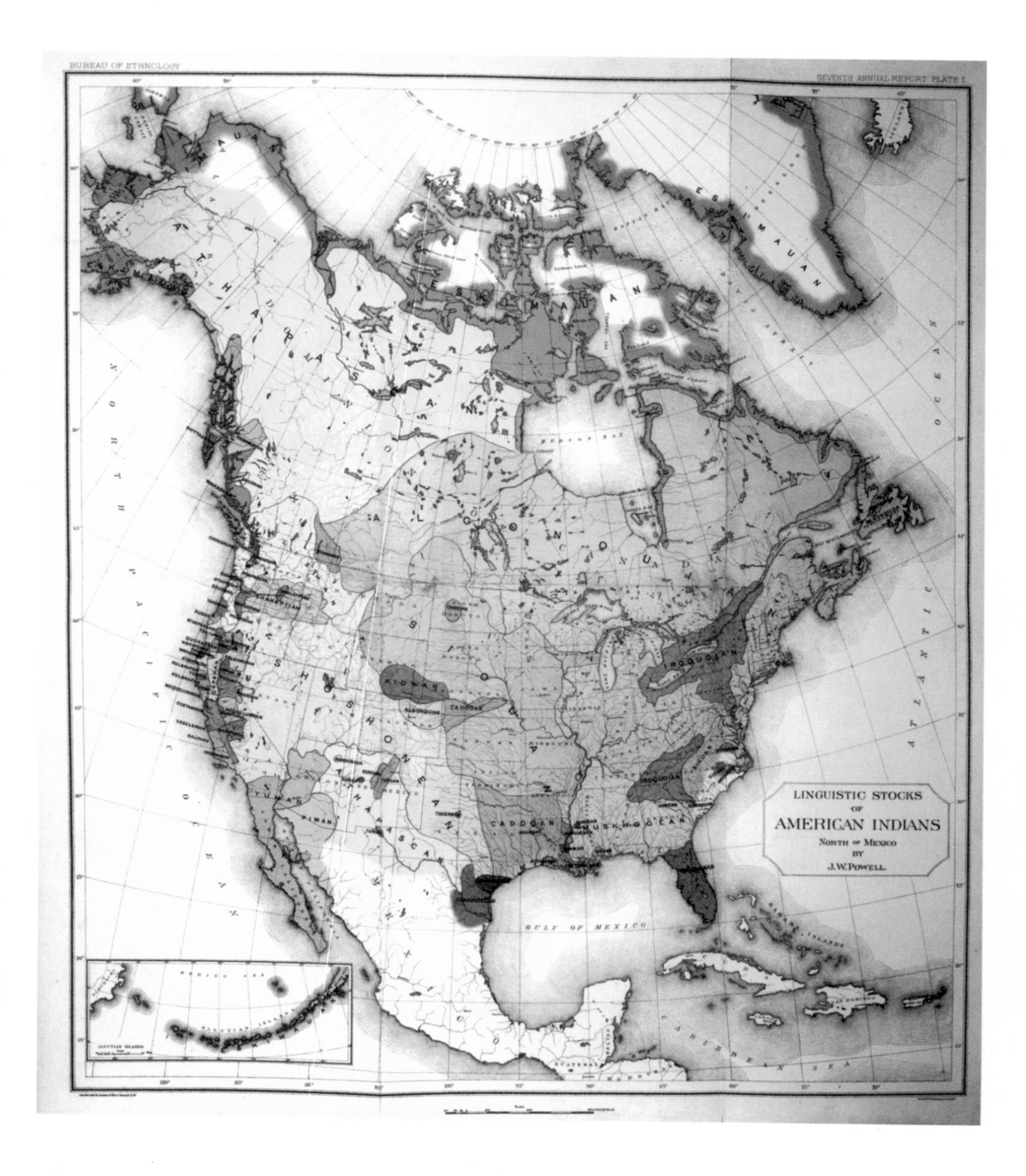

Spanish Imperium

New Spain, the earliest European empire in America, began with colonies in the Caribbean, and developed rapidly after Hernán Cortés encountered the rich civilization of the Aztecs in Central America in 1519. Motivated by the desire for more wealth, Spanish explorers pushed north into what became the south-western United States, eventually claiming the entire western half of the continent. Failing to find the fabled golden cities of Cíbola or the rich province of Quivira, the Spaniards established Catholic missions, which defined the Spanish presence on the frontier by the seventeenth century. Threats posed by the Indians, the French, and the Russians prompted Spain to add a series of presidios (fortified settlements) to protect the northern reaches of their empire near the end of the seventeenth century. By the late eighteenth century, the Spanish crown no longer sought to expand its American empire.

Spanish cartography reflected the evolving priorities of the empire. Sixteenth- and seventeenth-century maps furthered exploration and settlement; in the eighteenth century, more comprehensive surveys facilitated colonial government and defined the boundaries of farms and ranches. The Spanish crown controlled the administration of the empire, and most maps were manuscripts made for official use; as a security measure, they were prohibited from circulating. Because so few Spanish colonial maps of the American West were printed for broad distribution, they remain very rare.

When Mexico gained independence in 1821, California, New Mexico, and Texas became its northern provinces. The most distinctive cartographic expressions of this period were the *diseños*, maps submitted with land grant applications. To encourage the colonization of sparsely populated areas, the Mexican government offered large land grants to homesteaders, including Americans. Many settlers sketched their own *diseños*, producing lively vernacular renderings of the parcels they desired. After the territories became part of the United States, these Mexican-granted *ranchos* continued to appear on maps and shape settlement patterns. When the United States government later surveyed and divided the land, cartographers drew the new grid around the boundaries of the old *ranchos*.

1.8

Fortifying Spanish Borderlands

CREATORS: Don José de Azlor (draftsman), Sylverio (engraver)

TITLE: *Plan del Presidio de San Antonio de Bejar de la Provincia de Texas (Plan of the Presidio of San Antonio of Bejar of the Province of Texas)*, in Juan Antonio de la Peña, *Derrotero de la Expedicion en la Provincia de los Texas, Nuevo Reyno de Philipinas...* (Diary of the Expedition Made to the Province of Texas, New Kingdom of the Philippines...) (Mexico: Juan Francisco de Ortega Bonilla, 1722)

SIZE: 11.5 x 8.25 inches

LOCATION: Newberry vault Graff 3242
[not on display]

CREDIT: Gift of Everett D. Graff

In his *Derrotero de la Expedicion en la Provincia de los Texas*, Juan Antonio de la Peña recounts the 1721 expedition of Don José de Azlor, the Spanish Marques de San Miguel de Aguayo, to secure Texas in the wake of hostilities between France and Spain. In 1719 seven Frenchmen invaded Texas from Louisiana, prompting panicked residents of the six easternmost Spanish missions to flee to San Antonio. Aguayo, appointed Governor by the Viceroy of New Spain, organized a force of 500 men and 4000 horses to recapture the territory. His campaign was larger and more successful than any other Spanish expedition into Texas, securing Spain's control of the territory for the next one hundred and fifteen years. Aguayo founded or re-founded ten missions and four presidios, and suggested that Spain's hold could be strengthened by settling families in the territory.

Peña, a chaplain accompanying the campaign, penned his account in the form of a diary. His 1722 publication was the first book devoted wholly to the province of Texas. The volume's four plans document the presidios that resulted from the expedition. At the time, the area above the Medina River was sometimes referred to as the "New Philippines" or "New Kingdom of the Philippines."

Aguayo configured the presidio of San Antonio as a square, enabling a small number of men to defend it from two opposing corner bulwarks. The main entrance was along the south edge, opposite the church, which abutted the north wall. The bulwarks incorporated quarters for the Captain and other officers, while additional quarters lined the walls inside the complex. Dotted lines mark the corrals assigned to each dwelling. The drawing illustrates the local fauna as well as the presidio's strategic site. Aguayo located the fortress 30 varas (28 yards) from the San Pedro River and 200 varas (185 yards) from the San Antonio River. It made sense to situate the complex closer to the San Pedro, because its banks had few trees to impede defense. Aguayo also ordered the construction of an irrigation ditch from the San Pedro River to the presidio, enabling soldiers to grow corn on the land surrounding the complex.

References: Buckley 1911; Cumming 1974; Santos 1981; Wagner 1937b

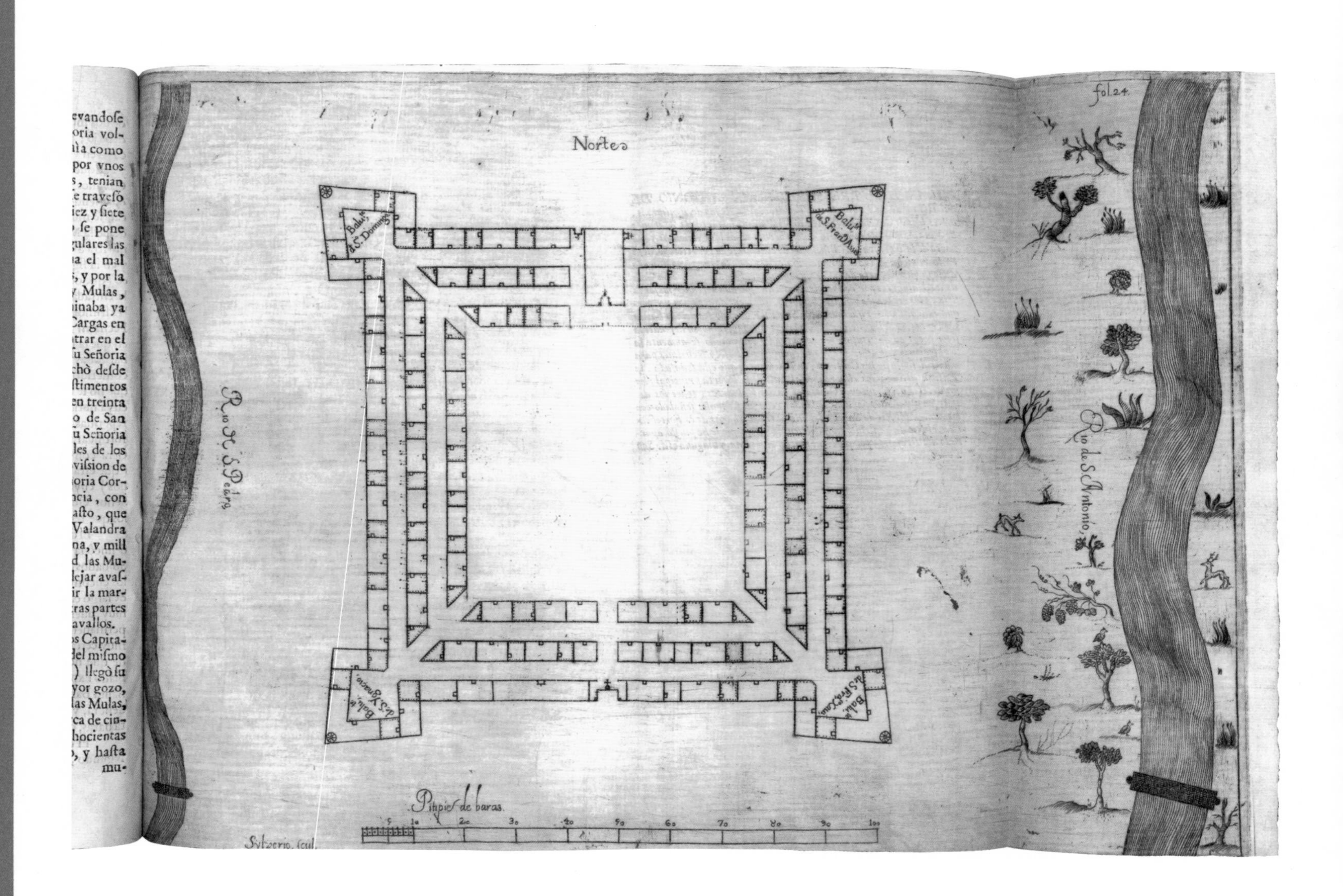

1.9

Surveying Northern New Spain

CREATOR: Francisco Álvarez Barreiro

TITLE: *Plano corographico è hydrographico, de las provincias de el Nuevo Mexico, Sonora, Ostimuri, Sinaloa, Culiacan, Nueba Vizcaya Nayarit, Nuevo Reyno de Leon, Nueva Extremadura o Coaguila, y la del Nuevo Reyno de Phipinaes, provincia de los Thefas...* (*Chorographic and hydrographic plan of the provinces of New Mexico, Sonora, Ostimuri, Sinaloa, Culiacan, New Vizcaya Nayarit, New Kingdom of Leon, New Extremadura or Coaguila, and of the New Kingdom of the Philippines, province of Texas*), 1728. Watercolor on parchment

SIZE: 20 x 35 inches

LOCATION: Hispanic Society of America, New York, NY
[not on display]

In his 1729 manuscript map, Francisco Álvarez Barreiro presented the first comprehensive depiction of the northern provinces of New Spain. He produced this general map, along with five specialized maps of individual provinces, to supplement the official report of Brigadier Don Pedro de Rivera about his expedition of 1724–1728. The Spanish viceroy had sent Rivera, then governor of Tlaxcala, to inspect the northernmost presidios and missions with an eye toward cutting administrative costs. Despite the French invasion that prompted Aguayo's 1721 expedition (see Item 1.8) into Texas, Rivera perceived no danger and recommended a reduction in the frontier militia. He also penned observations about the geography, ethnography, and economic resources peculiar to each province.

Barreiro, an experienced cartographer and engineer, provided the first systematic geographical survey of the region. He drew the boundaries of each province, calculated their latitude and longitude, and sketched the topography. In his descriptive account accompanying the maps, Barreiro explained that in addition to "all his majesty's dominions," the general map included "all the land that belongs to and is peopled by pagan nations." The general map, executed in watercolor, is enlivened by scores of miniature icons, representing mountains, forests, native settlements, and Spanish missions. The map also features what may be the earliest cartographic reference to the legend that the Aztecs migrated to Mexico from the vicinity of a lake at the headwaters of the Azul, a branch of the Gila River. A note next to the lake reads "Teguayo Lake or Blue Lake from where the Mexican Indians left with their Prince to found Mexico."

References: Cohen 2002; Thrower 1986; Wagner 1937b

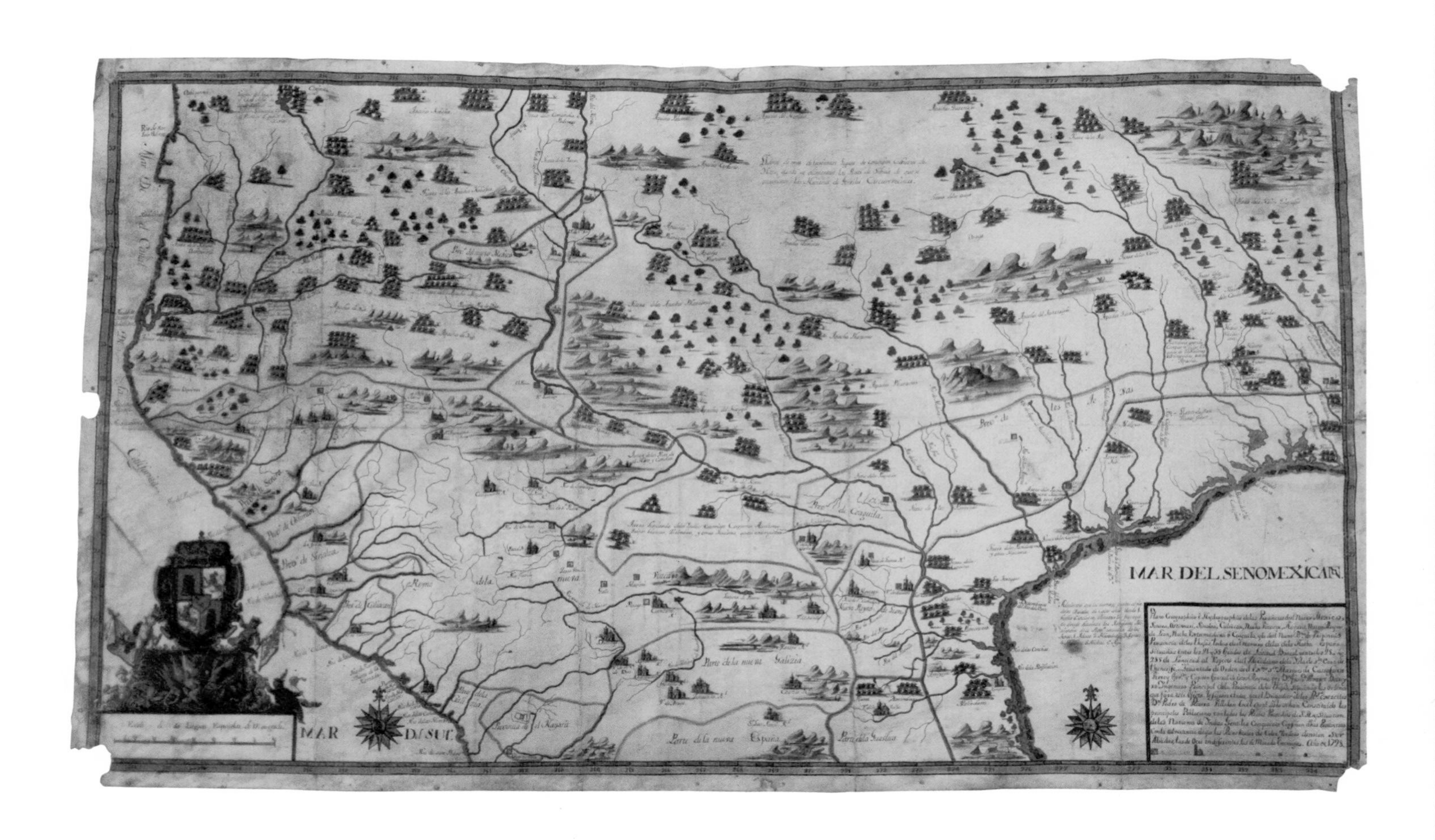

1.10

San Antonio Before the Battle of the Alamo

CREATORS: William Grattan (with annotations by William Bollaert)

TITLE: *Town of San Antonio or Bejar,* 1836. Pen and ink and pencil on paper

SIZE: 17.6 x 15.6 inches

LOCATION: Newberry vault oversize Ayer Art Bollaert sheet 30a

CREDIT: Gift of Edward E. Ayer

San Antonio proved the most durable Texas outpost of the Spanish empire. In 1718, the colonial government founded the Presidio of San Antonio de Béjar in tandem with the Franciscan Mission of San Antonio de Valero, to the east across the San Antonio River. To enhance security, the Spanish crown recruited settlers from the Canary Islands, who established the adjacent village of San Fernando de Béjar in 1731. The arrangement of the settlement exemplified the Spanish regulations for planning new towns, codified in the 1573 Laws of the Indies. The public plaza, the most distinctive feature of Spanish towns, was located at the grid's center.

Grattan's map, drawn as Texas battled Mexico for its independence, highlights military and political sites, including the government offices, jail, barracks, magazine, and infantry and cavalry quarters. The mission (marked "6" on the map) had been dissolved in 1793, and the building converted to a garrison known as the Alamo. It is possible that Grattan sketched the map to aid the Anglo-Texans' siege of San Antonio in November and December of 1835. Grattan's inclusion of the homes of two Mexican officers, General Martín Perfecto de Cos and Colonel Domingo de Ugartechea, who left San Antonio during the 1835 siege, suggests that he made the map before they departed. His schematic rendering of the Alamo itself also intimates that he drew the plan before the Mexicans recaptured the building on March 6, 1836. After the legendary battle, when the Alamo became a symbol of the Texans' sacrifice for their cause, the structure was usually represented more prominently and in greater detail.

William Bollaert, an English traveler, acquired the map during his 1842–1844 stay in Texas. The map's execution on transparent paper, in ink over pencil lines, indicates that Grattan or Bollaert probably traced it from another source. Bollaert may have labeled the mission "the Alamo" to emphasize its symbolic significance after the battle. The Englishman was fascinated by the decaying, picturesque structure, describing it in his journal as a "sacred pile of ruin." Bollaert wrote on September 20, 1843, that he was "going to the Alamo to make sketches." He may have used the map as a guide for this trip. At some point Bollaert seems to have used the folded map as scrap paper, adding the figures and notes about cotton exports at the right.

References: Bollaert 1956; Nelson 1998; Reps 1965; Schoelwer 1985

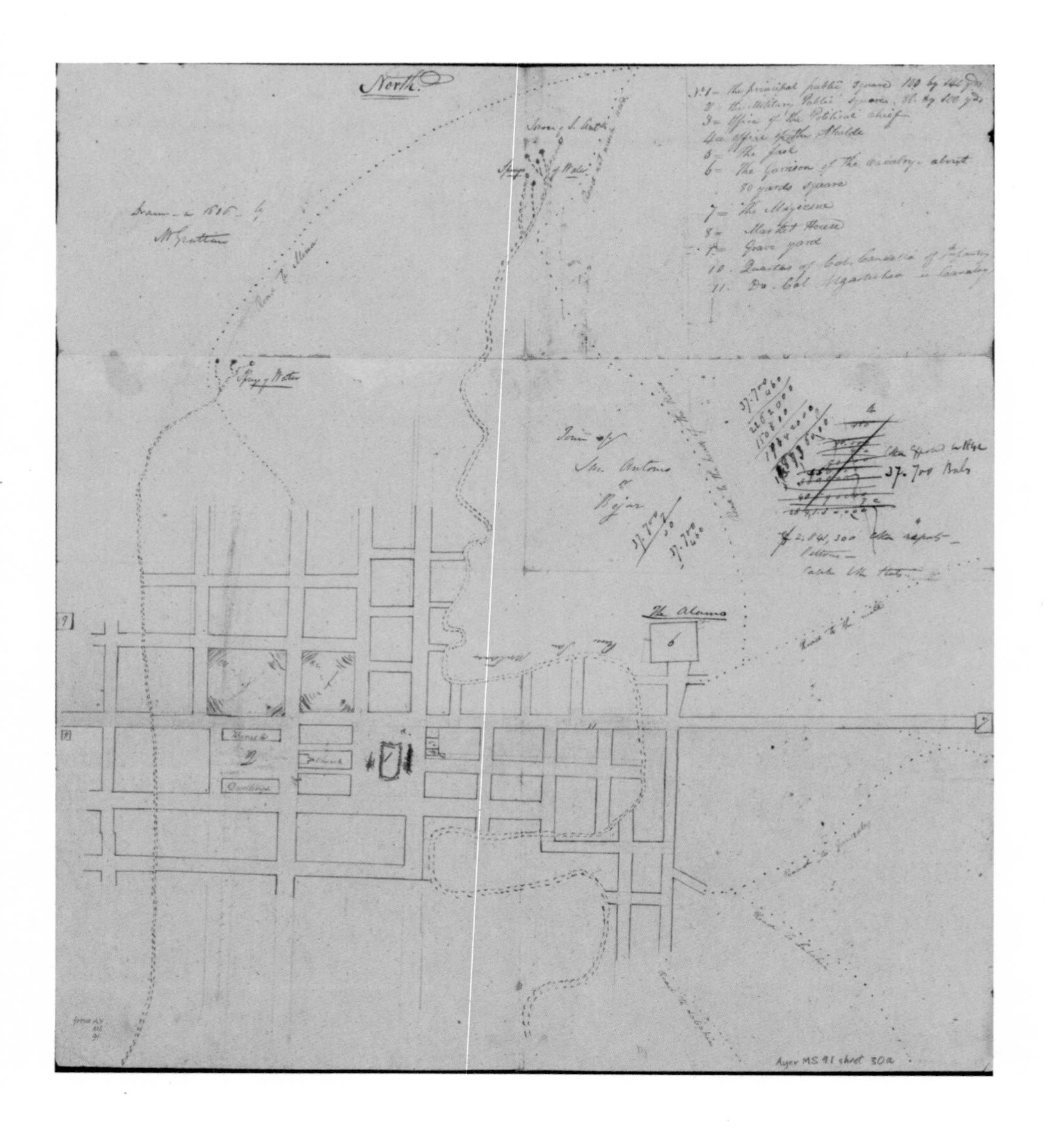

French Domain

Among the European empires vying for position on the continent in the colonial era, the French were the most energetic explorers and cartographers of lands that would become the American West. By 1498, French, English, Portuguese, and Spanish fishermen had discovered the rich cod banks off Newfoundland. French explorer Jacques Cartier made three expeditions between 1534 and 1542 in search of gold and a route to Asia. Cartier was the first European to navigate the St. Lawrence River, and his charts became the waterway's first maps. Permanent settlement began when Samuel de Champlain established Quebec in 1608. Champlain governed the colony for 25 years, forging alliances with Native Americans and importing Catholic missionaries to convert them. Champlain's maps depicted the interior features of the continent for the first time, creating the basis for France's territorial claims.

New France grew more rapidly after the crown took direct control of the colony in 1663 and appointed Jean Talon as governor. Eager to halt British expansion westward, Talon dispatched first Louis Jolliet and Father Jacques Marquette, and subsequently René-Robert Cavelier de La Salle, to chart the Mississippi Valley, and Daniel Greysolon, Sieur de Dulhut (Duluth) to explore the land west of Lake Superior.

The fur trade, soon the economic heart of New France, shaped its development. The antagonism between fur trapping (which required wilderness) and settlement encouraged the restriction of towns to a narrow crescent along the St. Lawrence, Great Lakes, and Mississippi. Because territorial boundaries east and west of the Mississippi were vague, France's empire overlapped with British and Spanish claims.

Tantalized by the storied wealth of the Spanish colonies, King Louis XIV accepted the proposal of Pierre Le Moyne, Sieur d'Iberville, for a Mississippi Valley colony. At first prosperity proved elusive, as Louisiana struggled to overcome political divisions, crop failures, and natural disasters. By the late 1730s, Louisiana's economy was stable and self-sufficient, supported by smuggling and low-level plantation production. Because the colony did not generate profits for the crown, French interest faltered. At the end of the French and Indian War in 1763, France ceded all of her mainland possessions to England and Spain.

French colonization in the American West yielded significant contributions to cartography. British and American mapmakers and consumers continued to rely for many years on the cartographic outlines established by the French.

1.11

The Cartography of Adventure

CREATOR: Luis Armand de Lom d'Arce, Baron de Lahontan

TITLE: *Carte de la Riviere Longue* (*Map of the Long River*), from *New Voyages to North-America: Containing an Account of the Several Nations of that Vast Continent …* (London: Printed for H. Bonwicke, T. Goodwin, M. Wotton, B. Tooke, and S. Manship, 1703), vol. 2.

SIZE: 14.25 x 9.5 inches

LOCATION: Newberry Graff 2364
[not on display]

CREDIT: Gift of Edward D. Graff

Louis Armand de Lom d'Arce, Baron de Lahontan (1666–c. 1715), came to Canada with the French colonial army in 1683, claiming to have explored western New France with his detachment of soldiers and five Indians between September 1688 and May 1689. Although much of the *New Voyages to North-America* that he published in 1703 is factual, Lahontan enlivened his romantic narrative and accompanying map with imaginary details, helping to shape popular European perceptions of the American frontier.

Lahontan used a dotted line to divide his map into two sections. To the east is his own diagram of the Great Lakes, Mississippi River, and Long River. Some scholars dismiss the Long River as imaginary; others recognize it as a distortion of the Minnesota River. The island-filled lakes could be condensed images of the series of long lakes into which the Minnesota River broadens. To the west is an area Lahontan claimed to have explored but actually did not reach. He recounted meeting a group of Gnacsitares (probably Siouian people), accompanied by four Mozeemlek slaves, on his return trip. The Mozeemlek offered "a Description of their Country, which the Gnacsitares represented by way of a map on Deer Skin." On this map, the hills at the far left most likely denote the eastern slope of the Coteau des Prairies, in northeastern South Dakota. The Gnacsitare mapmaker probably exaggerated their prominence because the hills signaled a change in environment, native population, and resources. Seen from the east, the hills marked the transition to an area where bison flourished.

Lahontan illustrated the map with the material culture of his fancifully named informants. He noted that the boat and long house of the Tahuglauk were based on Indian bark drawings. The medals to the right resemble the throat armor worn by some native peoples.

Although the map's most incredible features probably originated in Lahontan's imagination, some inaccuracies may have resulted from difficulties of communication, as Lahontan and the Native Americans he encountered did not speak the same languages. Throughout the first half of the eighteenth century, many respected cartographers accepted and reproduced Lahontan's assertions on their own maps.

References: Buisseret 1991; Cumming 1974; Lahontan 1703; Lewis 1998a; Lewis 1998b; Wheat 1957–1963

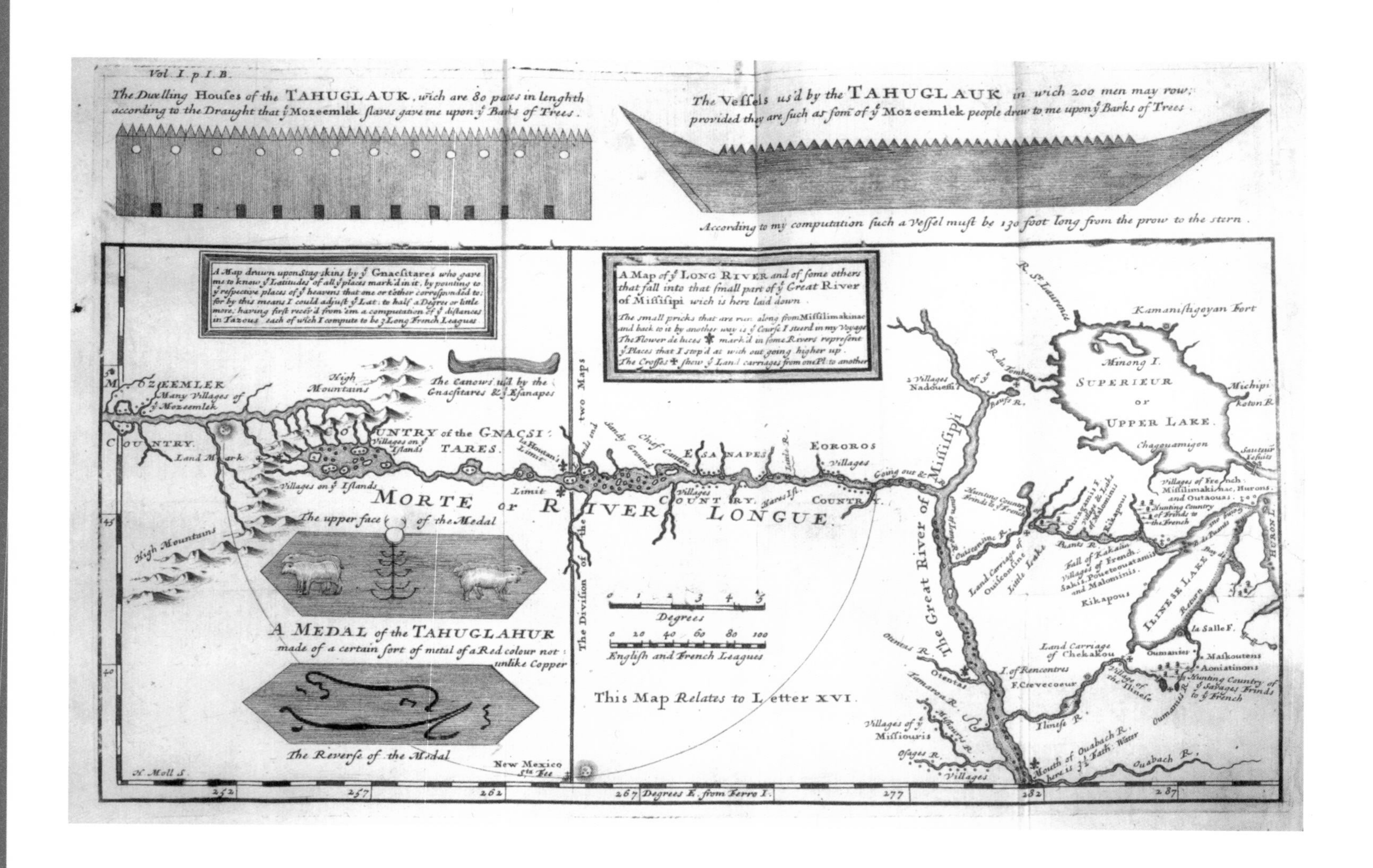

1.12

Charting Louisiana

CREATORS: Guillaume Delisle (cartographer), Académie Royale des Sciences (Paris, France) (publisher)

TITLE: *Carte de la Louisiane* (*Map of Louisiana*) (Paris: Académie Royal des Sciences, 1718)

SIZE: 22 x 26 inches

LOCATION: Newberry Map 4F G3700 1718 L5 (catalog image); The MacLean Collection (exhibit)

Summarizing the state of French geographic knowledge at the height of the nation's power in North America, Guillaume Delisle's (1675–1726) *Carte de la Louisiane* was perhaps the most influential map of the Mississippi Valley. A pioneer of scientific cartography, Delisle used astronomical observations to plot longitude and synthesized a wealth of primary sources. Although his maps were commercial products, they reflected French cartography's close ties to the state. Delisle never visited America, but his connection to the French court gave him access to the manuscript maps and geographical accounts in the Dépôt des Fortifications et Plans, the planning bureau for the colonies. These documents incorporated knowledge gleaned from Native Americans by French military officers, explorers and traders. Delisle obtained additional information from at least one missionary.

Carte de la Louisiane offered one of the first accurate depictions of the lower reaches of the Mississippi, Arkansas, and Red Rivers, and the surrounding Gulf coast, and was the first modern map to trace Hernando de Soto's route. Delisle added New Orleans, founded the year the map was first issued, to later editions. His depiction of areas beyond French settlement—the upper reaches of the rivers, Lake Michigan, the Rocky Mountains, and Florida—are less precise.

Delisle aroused the ire of France's imperial rivals by extending Louisiana from the Rio Grande to the Appalachian Mountains, encompassing lands claimed by the Spanish to the west and the English to the east. His map further provoked the British by declaring that the French first discovered and settled "Caroline," naming Charlestown for their king. English mapmakers quickly countered by aggressively asserting Britain's territorial claims.

References: Buisseret 1991; Cumming 1974; Lagarde 1989; Lemmon, Magill, and Wiese 2003; Pedley 2005; Pelletier 1983; Schwartz and Taliaferro 1984; Wheat 1957–1963

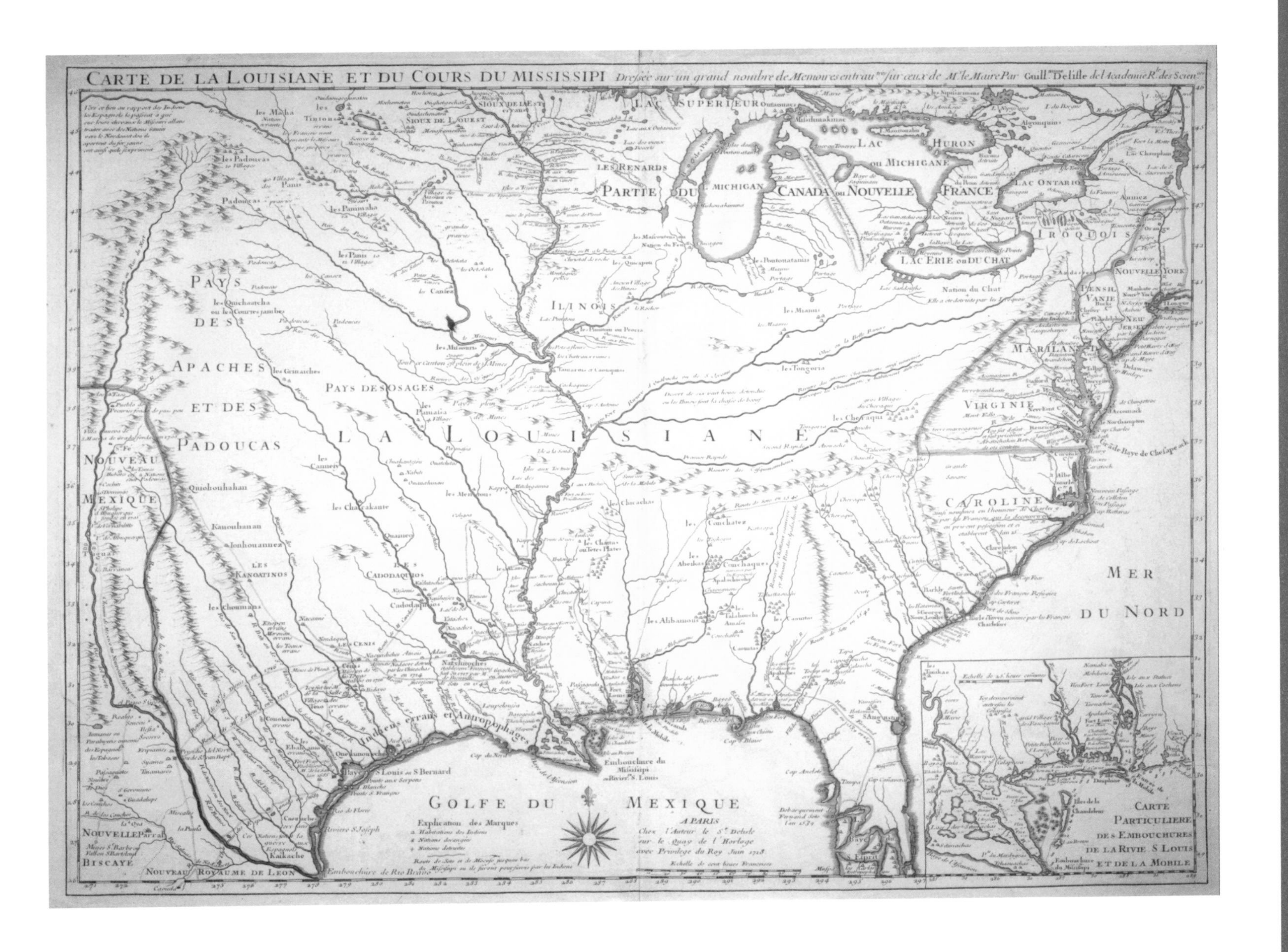

1.13

Selling Louisiana

CREATORS: Nicolas de Fer, Jacques-François Bénard (cartographers, publishers)

TITLE: *Le Cours du Missisipi ou de St. Louis fameuse riviere d' Amerique septentrionale aux environs de laquelle se trouve le pais appelle Louisiane...* (*The Course of the Mississippi or St. Louis, Famous River of North America Surrounding Which one Finds the Country called Louisiana*) (Paris: Chez J. F. Benard, 1718)

SIZE: 37.6 x 24.4 inches

LOCATION: Newberry Map 6F G4042 M5 1718 F4
[not on display]

In contrast to Delisle's scientific view of Louisiana, and in the same year, Nicolas de Fer (1646–1720) offered a more decorative and promotional image of the French colony. De Fer was appointed Geographer to the French King in 1691. The son of a Paris mapseller, de Fer took over the family business in 1687 and developed it into a successful map publishing firm. His sons-in-law Jacques-François Bénard and Guillaume Danet continued the business after de Fer's death.

De Fer produced *Le Cours du Missisipi* for the *Compagnie d'Occident* (Company of the West, renamed Company of the Indies in 1719), formed to develop a French colony in the Mississippi Valley. King Louis XIV sought private capital to build the colony after his extravagant spending depleted the French treasury. The king first granted exclusive trading privileges to financier Antoine Crozat in 1712; when Crozat relinquished his charter in 1717, Scotsman John Law stepped in, aggressively marketing shares in his Company of the West throughout Europe. Stock prices rose dramatically until the "Mississippi Bubble" burst in 1721.

De Fer's attractive map, with the firm's coat of arms placed alongside those of Quebec at the top of the sheet, provided effective advertising for Law's company. De Fer portrayed the valley as a resource-rich territory ripe for development. The land is teeming with Native Americans, who seem to thrive on the abundant game and plant life. The Indians appear to move easily through the interior on foot and in canoes, paddling from one end of the territory to the other through a network of lakes, rivers, and the Gulf of Mexico. The landscape is dotted with settlements, from Detroit and Chicago along the Great Lakes to Santa Fe and Taos along the Rio del Norte (upper Rio Grande) in the southwest. The native humans and animals seem exotic but not dangerous, enticing viewers to come experience the landscape for themselves.

The map shown here comprises the western half of the map de Fer fabricated to illustrate the entirety of France's interests in North America. He made the large map by fitting together four separately created sheets, each based on the surveys and manuscript maps prepared by members of d'Iberville's exploring party. De Fer minimized the differences in scale and area among the four maps by trimming away their individual titles and borders. Along the top, he included a detailed inset of the Gulf of Mexico.

References: Cumming 1998; Karrow and Buisseret 1984; Verner 1979

1.14

A Military Officer's Sketch of New Orleans

CREATOR: Jean François Benjamin Dumont de Montigny

TITLE: *Plan de la Nouvelle Orleans, Ville capitalle de la Louissianne* (*Plan of New Orleans, Capital City of Louisiana*), from *Mémoire de Lxx Dxx officier ingénieur, contenant les evenements qui se sont passés à la Louisiane depuis 1715 jusqu'à present* (*Memoir of Lxx Dxx, officer engineer, containing the events that have taken place in Louisiana from 1715 to the present*). 1747. Watercolor on paper

SIZE: 13.6 x 18.2 inches

LOCATION: Newberry Ayer MS 257, sheet 7

CREDIT: Gift of Edward E. Ayer

Jean François Benjamin Dumont de Montigny (1696–after 1754), a Paris-born military officer, keenly observed the daily life and environment of Native Americans and colonists during his extensive travels through New France. Dumont spent two years in Quebec and more than 17 years in Louisiana. He helped build numerous French forts, serving as a draftsman, engineer, and plantation manager, as well as a soldier. Ten years after his 1737 return to France, Dumont penned a memoir of his American adventures, copiously illustrated with watercolor drawings and maps.

Dumont pictured New Orleans as he remembered it appearing in the 1720s and 1730s, visualizing the environmental and social challenges facing the new settlement. The city had been founded in 1718 by Jean Baptiste le Moyne, Sieur de Bienville (d' Iberville's younger brother), as part of the scheme of John Law's company to colonize Louisiana (see Item 1.13). After a hurricane destroyed the initial settlement in 1722, the company hired engineer Adrien de Pauger to survey and plot the town. Pauger's rational grid exemplified eighteenth-century urban planning.

Dumont contrasted the rectilinear plan with the curving natural forms of the surrounding trees, rivers, and ponds. The low-lying swampy land risked flooding, was difficult to drain, and offered an unstable foundation for construction. Impressed by Pauger's rationalization of space, Dumont reproduced the design of the town on three scales. The drawing at the center diagrams the overall street grid. The large square to the right enlarges four blocks to show the division of the land into symmetrical long lots. The rectangle at the far left portrays a typical domestic lot, with the house and outbuildings facing the street and a subsistence garden at the rear. A high fence encloses the lot, to protect the property from marauding animals and provide a sense of security. The small square to the left of the central grid indicates Bienville's country estate. The bar extending northwest from the river represents a canal that was never built. Further upriver (to the left) Dumont sketched the Jesuit plantation.

References: Buisseret 2007; Dawdy 2008; Dumont de Montigny 2008; Reps 1965

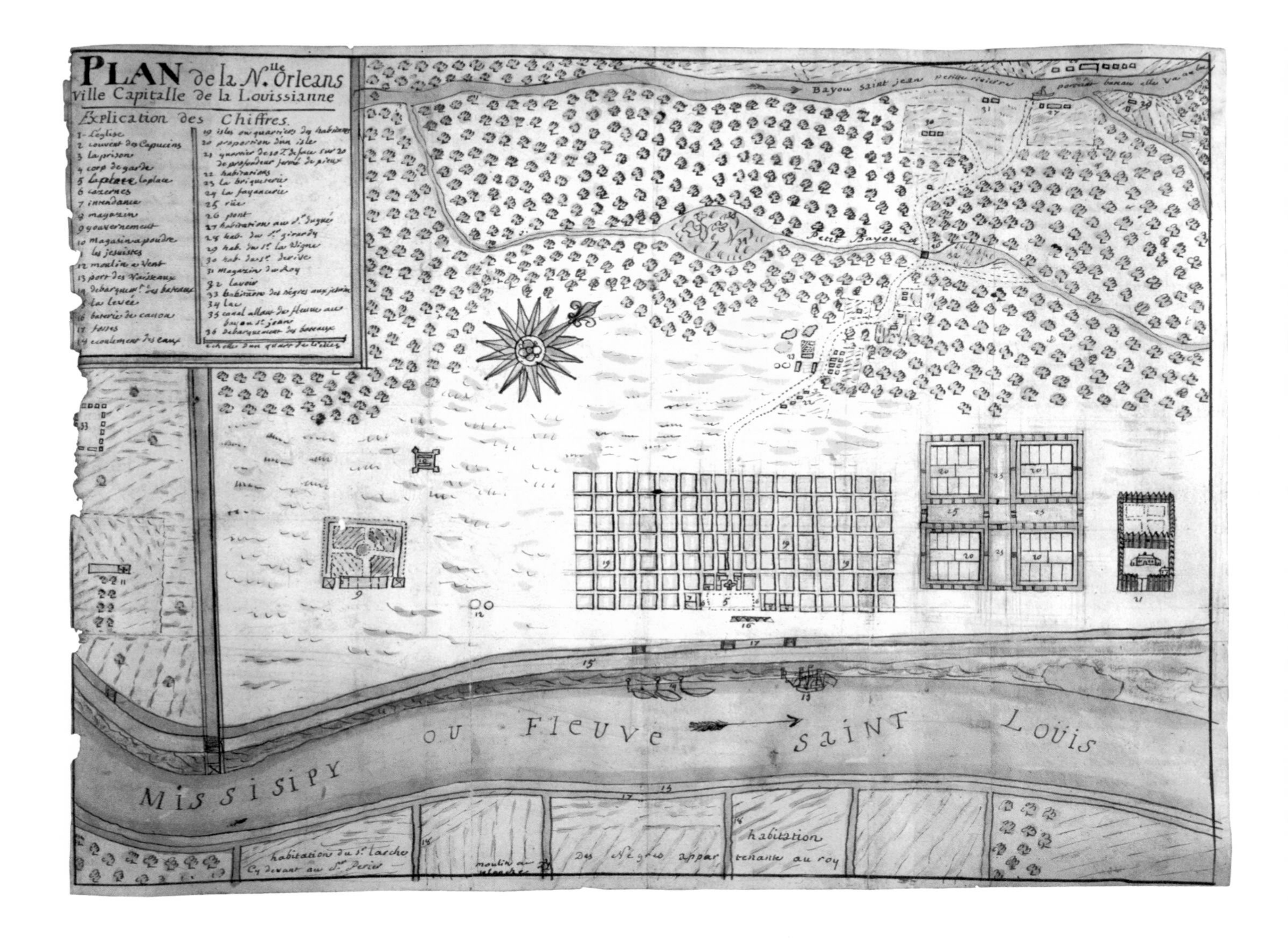

British Territory

The British mapping of North America followed a pattern distinct from that of the Spanish and French. Because the colonies were concentrated along the Atlantic seaboard, early English maps focused on the coast rather than the interior. The colonies, founded by private joint-stock companies, were governed by independent administrations in America, overseen by lower-level bureaucrats in London. As neither group had the resources to commission general maps, commercial cartographers supplied them. For more specialized needs, such as determining the best location for forts or negotiating boundary disputes, officials ordered manuscript maps.

British interest in maps of the interior increased as competition with the French over control of the Trans-Allegheny region escalated. The royal charters of Virginia (1609) and New England (1620) granted the colonies land "from Sea to Sea," and the English regarded this territory as theirs. When victory in the French and Indian War gave the British Empire control of nearly all of North America east of the Mississippi River, the government sought geographic information about the area. Because French cartographers had produced the most detailed and reliable maps of the interior, London mapsellers met much of this new demand with French maps or British maps based on French sources.

1.15

North America from a British Vantage Point

CREATORS: John Mitchell (cartographer), Thomas Kitchin (engraver), Andrew Millar (publisher)

TITLE: *A Map of British and French dominions in North America, with the roads, distances, limits, and extent of the settlements...* (London: Andrew Millar, 1755)

SIZE: 52.4 x 76 inches

LOCATION: Newberry vault Ayer 133 M66 1775 (catalog reproduction); The MacLean Collection (exhibit)

The rising political significance of the Western colonial territories inspired *A Map of the British and French Dominions in North America*. George Montagu-Dunk, second earl of Halifax, recognized the fertile, mineral-rich land between the Appalachian Mountains and the Mississippi River as the linchpin to the larger balance of power between England and France. As President of the Board of Trade and Plantations (the agency responsible for the colonies), Halifax commissioned the map to illustrate French encroachments on territory claimed by the British. He hired John Mitchell (1711–1768) to synthesize geographical information in the Board's archives. Mitchell, a physician recently arrived in London from Virginia, had scant mapmaking experience but considerable knowledge of the colonies. The manuscript map he produced in 1750 revealed gaps in the Board's archives, convincing Halifax to order an improved version. Halifax directed the colonial governors to supply updated maps and boundary descriptions, to which Mitchell added information from his own American contacts. He integrated these new details onto a base adapted from Guillaume Delisle's maps (see Item 1.12). Although produced decades earlier and reflecting a French perspective, Delisle's work still offered the best outline of North America. Mitchell refined the coastline by studying the observations of British Navy officers recorded in ships' logs.

The finished map was engraved by Thomas Kitchin, published by Andrew Millar in London, and first advertised for sale in the *Public Advertiser* on March 28, 1755. Because the map was costly, its buyers were primarily politically engaged members of the London elite, but inexpensive derivatives reached a broader audience. As tensions rose between England and France, the map rallied British support for the French and Indian War, the North American counterpart to the Seven Years War (1756–63) in Europe.

Mitchell conveyed the argument about England's territorial rights through the cartouche and the textual commentaries scattered across the map. The cartouche pictures two Iroquois, whose position below the title, royal crest, and British flag indicates their deference to British authority. The commentaries detail the Iroquois' historic occupation of lands around the Great Lakes and the Ohio River valley, which the British could now claim by regarding the Iroquois as dependents. Other texts assert British rights to land further south by prior exploration and settlement.

British and American diplomats used John Jay's copy of Mitchell's map to negotiate the Treaty of Paris in 1782 and determine the boundaries of the new United States. Created to assert British claims in North America, the map now helped to pronounce the end of their colonial reign.

References: Berkeley and Berkeley 1974; Edney 1997; Edney 2007; Edney 2008

1.16

George Washington, Emissary and Mapmaker

CREATOR: George Washington

TITLE: *Sketch map of the country traversed by Washington on his journey in 1753–54 from Cumberland, Maryland, to the site of Pittsburgh, Pa., and Fort le Boeuf near Lake Erie,* 1754. Brown ink with brown and black ink washes on paper

SIZE: 18.88 x 15 inches

LOCATION: Private collection

George Washington

George Washington in the Uniform of a British Colonial Colonel, 1772, by Charles Willson Peale.

CREDIT: Washington-Custis-Lee Collection, Washington and Lee University, Lexington, VA

Before he was a revolutionary leader and an American President, George Washington (1732–1799) was a land surveyor in Virginia and a soldier in the British Army. His sketch map of the Ohio River region illuminates his cartographic skills, his imperialist inclination, and his identification of future prosperity with the Western frontier.

Competition between England and France for the area surrounding the Forks of the Ohio River, where Pittsburgh now stands, prompted the map's production. According to its royal charter, Virginia extended to the Pacific. Planting tobacco generated wealth but exhausted the soil, sending colonists west for fresh land. In the 1740s, the Ohio Company, a land-speculating syndicate of Virginia planters, began eyeing the Forks area. The French likewise had ambitions to control the territory. After the Ohio Company built Red Stone Fort a mere 37 miles from the Forks in 1752, the French governor, Ange Duquesne de Menneville, Marquis Duquesne, moved quickly to establish four forts between Lake Erie and the Forks. Robert Dinwiddie, Virginia's lieutenant governor and an Ohio Company shareholder, saw the French forts as an attempt to halt Virginia's expansion. With permission from the crown, Dinwiddie sought to repel the French intruders. His first step was to send an emissary—twenty-one-year-old Washington—to the French officials at Fort Le Boeuf, demanding their withdrawal.

Washington's manuscript map traces the route of his westward journey from the Ohio Company's storehouse on the Potomac River to the French forts along the upper Ohio River. Based on his survey, Washington suggested a revised location for the Ohio Company's planned fort at the Forks. When Washington returned to Virginia with news that the French commander had refused to abandon the fort, Dinwiddie asked that he write an account of his expedition for publication. The governor sent the report, accompanied by Washington's manuscript map, to London, where it was immediately printed with an engraved map. The account convinced British officials that Dinwiddie should erect a fort at the Forks. Washington's knowledge of the area secured his appointment as lieutenant colonel of the regiment formed to build and guard the fort. In 1754, escalating tensions between the British and French over control of the Ohio Valley erupted into the French and Indian War.

References: Anderson 2000; Anderson 2005; Ellis 2004; Lewis 1993; Schwartz 1994

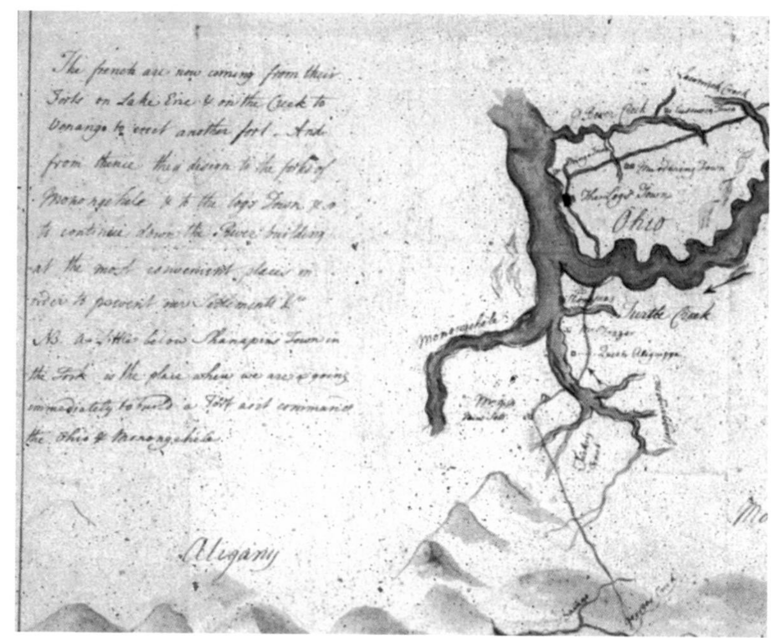

1.17

Picturing British Detroit

CREATOR: Lieutenant Edmund Henn

TITLE: *A View of Detroit. July 25th, 1794.* Pencil, pen and ink, and watercolor on paper

SIZE: 11.64 x 18.84 inches

LOCATION: Burton Historical Collection, Detroit Public Library, Detroit, MI [not on display]

The British gained control of Detroit from the French in 1760, occupying the outpost until 1796, when it passed belatedly to the United States as part of the Jay Treaty. Lieutenant Edmund Henn captured the topographic and social character of the late eighteenth-century British garrison in his view of the town from the southwest. Henn seems to have served at Detroit for much of 1792–96, when his unit, the 24th Regiment of Foot, was posted there.

The painting reveals the soldier's eye for strategic detail. Henn devoted much of his composition to the river, which enabled Detroit's military and commercial development. The large ships on the right are the Ottawa and the Chippewa, armed vessels built around 1792. In the background, older ships that had served in the Revolutionary War line the King's Wharf. The billowing smoke probably rises from the caulking pots, where shipwrights melted pitch to seal the ships' hulls. On shore, the town is protected by a stockade and defended from blockhouses. The largest structure, at the far left, is the officers' quarters. Through the open gate Henn provides a glimpse of the town's main thoroughfare, Saint Anne Street, parallel to the river.

Henn also inventoried Detroit's diverse population. Native Americans paddle the canoe near the bank; they may represent Ojibwa from the islands in Lake St. Clair or Wyandots from further down the river. The figure on the shore seems French, judging by his two-wheeled, Canadian-style cart. Detroit's most recent residents, the British, include the soldiers guarding the gateway and the civilians fishing in the foreground. An African slave paddles the fishing party's canoe.

Reference: Dunnigan 2001

Russian Country

Russia's colonization of America's northwest coast grew out of Tsar Peter the Great's expansionist ambitions. Europeans knew little about the northern Pacific Ocean until the tsar sponsored expeditions led by Russian Naval officers Vitus Bering, Aleksei Chirikov, and Mikhail S. Gvozdev in the early eighteenth century. The explorers brought back luxurious furs, which the Russians sold at a great profit in China.

The highly-desired furs inspired Russian merchant Grigorii I. Shelikhov to found a settlement on Kodiak Island off the Alaskan coast in 1784. The imperial government consolidated the trade in 1799 to form the Russian American Company, which became the main rival of the British Hudson's Bay Company in the global fur business. In the same year, the firm's chief manager, Alexander Baranov, moved their headquarters to New Archangel (or New Saint Michael's, now Sitka) in 1799. The new settlement served as the administrative capital of Russian America from 1808 until the transfer of Alaska to the United States in 1867. In 1812 Baranov founded Fort Ross, an agricultural outpost in California, which supplied Russian America with vegetables, fruit, and meat.

Unlike the other European powers in America, the Russians did not aspire to establish expanding colonies. The Russian population in America rarely exceeded 500 at any time, and natives outnumbered Europeans in every settlement. By the 1830s, the tsar forbade permanent emigration, allowing only those married to natives to remain for life. Moreover, the Russians expressed little interest in the continental interior, restricting their enterprises almost entirely to the coast. These colonial policies limited the demand for images of the region, making Russian maps, views, and atlases of America especially rare.

1.18

A Russian Community Takes Shape: New Archangel in 1842

CREATOR: Rostislav Grigor'evich Mashin

TITLE: *[Plan of Sitka]*, in Eduardo Blaschke, *Topographia medica portus Novi-Archangelscensis, sedis principalis coloniarum rossicarum in Septentrionali America [A Medical Topography of the Port of New Archangel]* (Petropoli: typis K. Wienhöberi et filii, 1842)

SIZE: 13 x 39.8 inches

LOCATION: Newberry vault Ayer 160.5 A3 B644 1842

CREDIT: Gift of Edward E. Ayer

Eduard Leonidovich Blaschke (1810–1878) became the chief physician for the Russian American Company shortly after he completed medical school, arriving in New Archangel (later Sitka) in 1835. He remained until 1840, when he became surgeon to the round-the-world vessel Nikolai I. In 1842, Blaschke published *A Medical Topography of the Port of New Archangel* in St. Petersburg. The inclusion of a plan of New Archangel followed from Blaschke's view that environmental factors caused diseases, and that treatment required knowledge of the local area. During his residence Blaschke surveyed the vicinity's physical and social geography, studying the weather, water, and soil, along with health patterns and modes of life.

Navy Lieutenant Rostislav Grigor'evich Mashin's 1838 map documents the extent of New Archangel. The chief manager's house appears at the center of the largest headland at the southwest, protected by a palisade and watch towers. The third house on this site, the structure was completed in 1837 and remained in place at the time of Alaska's transfer to the United States in 1867. Directly behind it are houses and barracks for company employees, warehouses, and the arsenal. Beyond are the admiralty buildings, officers' quarters, naval school, and shops—all large communal structures built of massive, hewn logs. The ship anchored in the water, at the far right of the diagram, no longer seaworthy, served as a warehouse. At the edge of the settlement stood the marketplace for the Tlingit Indians, whose houses are represented by the rectangles along the northwest shore (lower right).

St. Michael's Church (predecessor to the cathedral later built nearby) occupies the central peninsula, while the hospital dominates the northernmost promontory. To the rear are private dwellings and the chief manager's vegetable garden. Crosses mark the Russian cemetery, separated from the Tlingit settlement by a ridge. The island opposite the hospital included a warehouse for salted fish. The plots stretching along the coast to the northeast mark additional vegetable gardens and subterranean caves filled with ice.

References: Alekseev 1987; Andrews 1922; Black 2004; Pierce 1986; Pierce 1990b

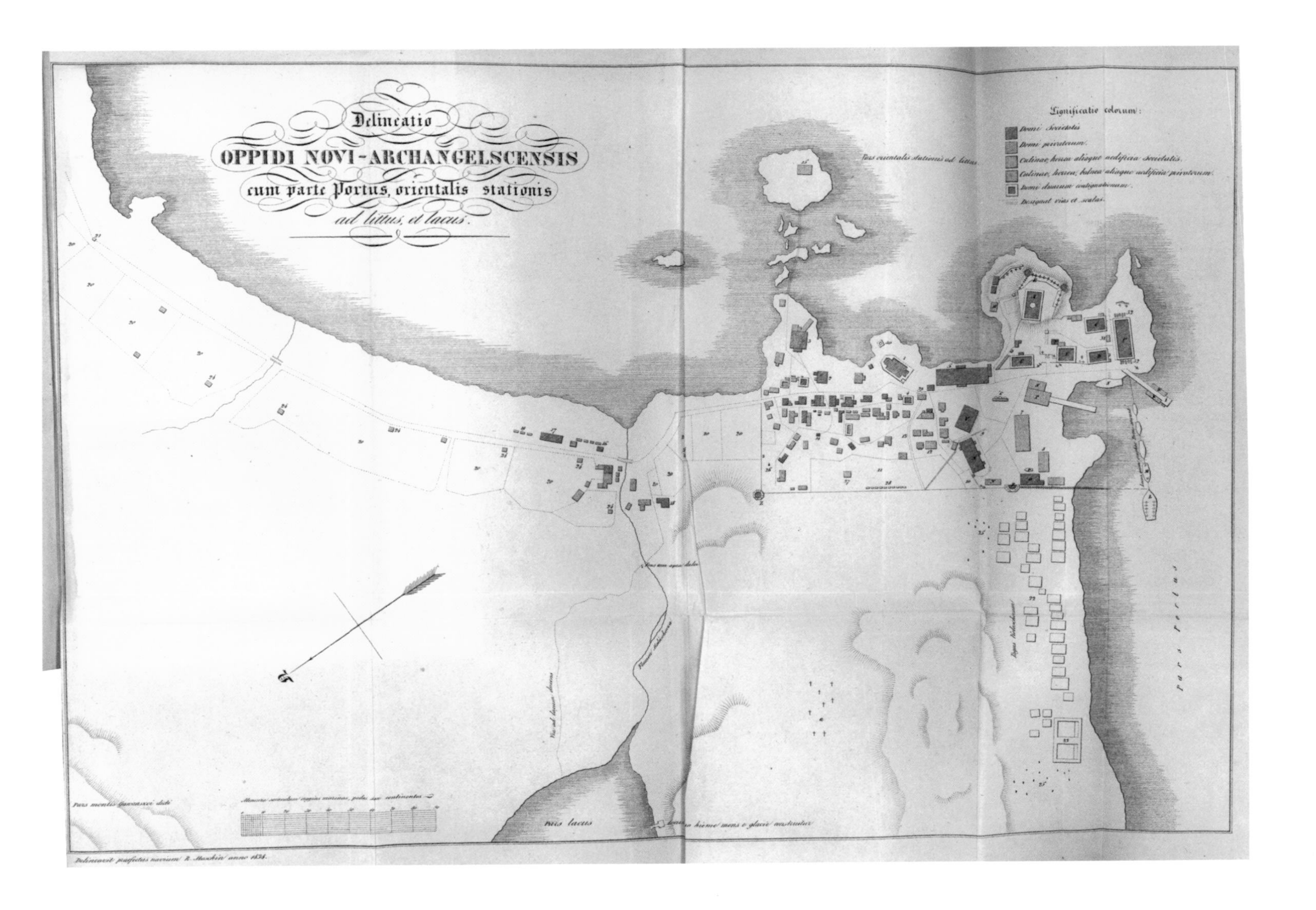

1.19

Charting Russian Waters

CREATORS: Mikhail Dmitrievich Teben'kov (compiler), Mikhail Kadin (draftsman), Grigorii Klimovich Terent'ev (engraver)

TITLE: *Karta Sievernoi chasti Sievernago Tikago Okeano* (*Map of the North Part of the North Pacific Ocean*), in Mikhail Dmitreivich Teben'kov, *Atlas sieverozapadnykh beregov Ameriki ot Beringova proliva do mysa Korrientes i ostrovov Aleutskikh s prisovokupleniem niekotorykh miest sieverovostochnago berega Azii.* ([Sanktpeterburg], 1852) [Atlas of the Northwest Coasts of America from Bering Strait to Cape Corrientes and the Aleutian Islands with Several Sheets on the Northeast Coast of Asia] ([St. Petersburg], 1852)

SIZE: 21 x 39 inches

LOCATION: Newberry vault oversize Graff 5016

CREDIT: Gift of Everett D. Graff

Mikhail Dmitrievich Teben'kov (1802–1872) spent twenty-five years in Alaska and the North Pacific as a naval officer, explorer, cartographer, and government administrator. When he was appointed chief manager of the Russian colonies in America in 1845, he determined to improve coastal navigation by compiling an updated hydrographic atlas.

For the atlas's 39 charts, Teben'kov synthesized information gathered during his own voyages, from published accounts, and from coastal sketches drawn by natives. To supplement this data, he directed subordinates to make further expeditions and examine ship logbooks. Mikhail Kadin, an Alaska-born pilot and drafting teacher, drew the charts. Grigorii Klimovich Terent'ev, a Russian clerk and mechanic, also born in Alaska, engraved the maps on copperplates. When Teben'kov's term as chief manager ended in 1850, he took the plates to St. Petersburg and supervised the printing of the atlas.

In the volume of *Hydrographic Notes* Teben'kov wrote to accompany the atlas, he identified his primary goal as the mathematical determination of locations along the entire American coast. The *Notes* lists the latitude and longitude of most major ports, capes, and bays. Teben'kov's general map shows the full scope of Russia's empire in the North Pacific, which briefly included outposts in California and in the Hawaiian Islands. The increasing compression of the territory toward the equator resulted from Teben'kov's use of a Mercator projection.

References: Andrews 1922; Black 2004; Falk 1990; Pierce 1990b; Teben'kov 1981

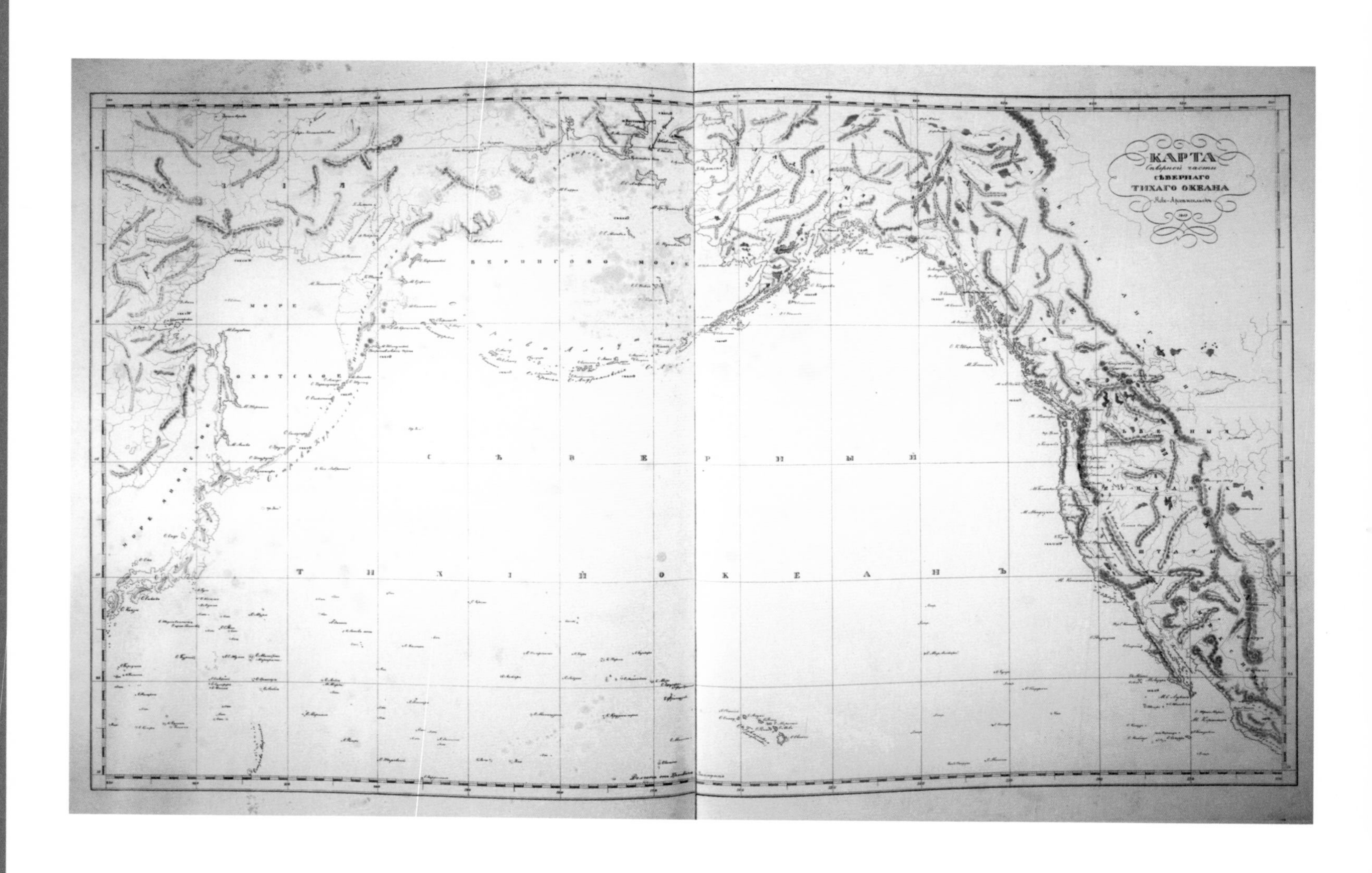

1.20

The Capital of Russian America: Sitka in 1868

CREATOR: A. De Stannicovsky

TITLE: [*Panorama of Sitka*], c. 1868. Possibly tempera on paper

SIZE: 13.25 x 39.75 inches

LOCATION: Newberry vault drawer MS Map6F G4374 S5A3 1868 S7

CREDIT: Gift of Anne Haffner

The artist A. de Stannicovsky pictured New Archangel (Sitka) at the end of its Russian heyday. The American flag flying prominently attests that the transfer of October 18, 1867, has taken place, while the Russian flag at half-mast could represent its ceremonial lowering. The artist may have achieved the elevated perspective by viewing the scene from the top of a ship's mast. As there is no record of Stannicovsky's residence in New Archangel, he may have been part of the crew on the ship that transported colonists back to Russia after the transfer.

Russia's decision to withdraw from Alaska was prompted in part by Britain's escalating naval power in the Pacific and the United States' westward expansion. For the Russian and native inhabitants of New Archangel, however, life was fairly good. On the eve of the transfer, trade and industry were thriving and the community supported five schools and three churches. Residents enjoyed steam baths, tea houses, public gardens, and social clubs.

Stannicovsky's jewel-like colors emphasize the town's imperial maturity. On the promontory at the left is the chief manager's residence, also represented on the plan in Blaschke's *Topographia*. Though he never lived there, the building became known as "Baranov's Castle" in recognition of his historic prominence. On the shore in front of the dwelling is a bath house, and behind it a two-story office building. The green spire and onion dome topped by Greek crosses near the center of the panorama identify St. Michael's Cathedral, consecrated in 1848 and designed in the typical Russian orthodox style. The only log cathedral ever built in America, it featured bells donated by the church in Moscow and twelve icons decorating the interior. The trees to the right of the cathedral mark the Russian tea garden. The sawmill, identified by its waterwheel and the logs scattered in front, attests to the area's rich timber resources. Nearby forests supplied planks for the shipbuilding industry, as well as for the town's buildings and sidewalks. The white building at the far right is the hospital, while the gray structure next to it served as the bishop's house and monastery.

When Secretary of State William Seward negotiated the purchase of Alaska's 586,412 square miles for $7 million, critics pronounced it "Seward's Folly." The discovery of gold and oil in the 1890s made the deal seem like a bargain.

References: Andrews 1922; Black 2004

Chicago Becomes a Place

These map details trace Chicago's evolution from a Native American place name ("Chekagou") on Coronelli's 1688 map into a gridded urban landscape on Rees's 1851 sheet; from a mere spot on a map of a barely understood continent to a city with its own identity and sense of purpose. Chicago gained its place on the national stage because it established itself as a transportation hub for raw materials, finished goods, and people. First by water and then by railroad, this transportation network linking the West to the East spurred the city's eventual development into a major center of cartographic publishing.

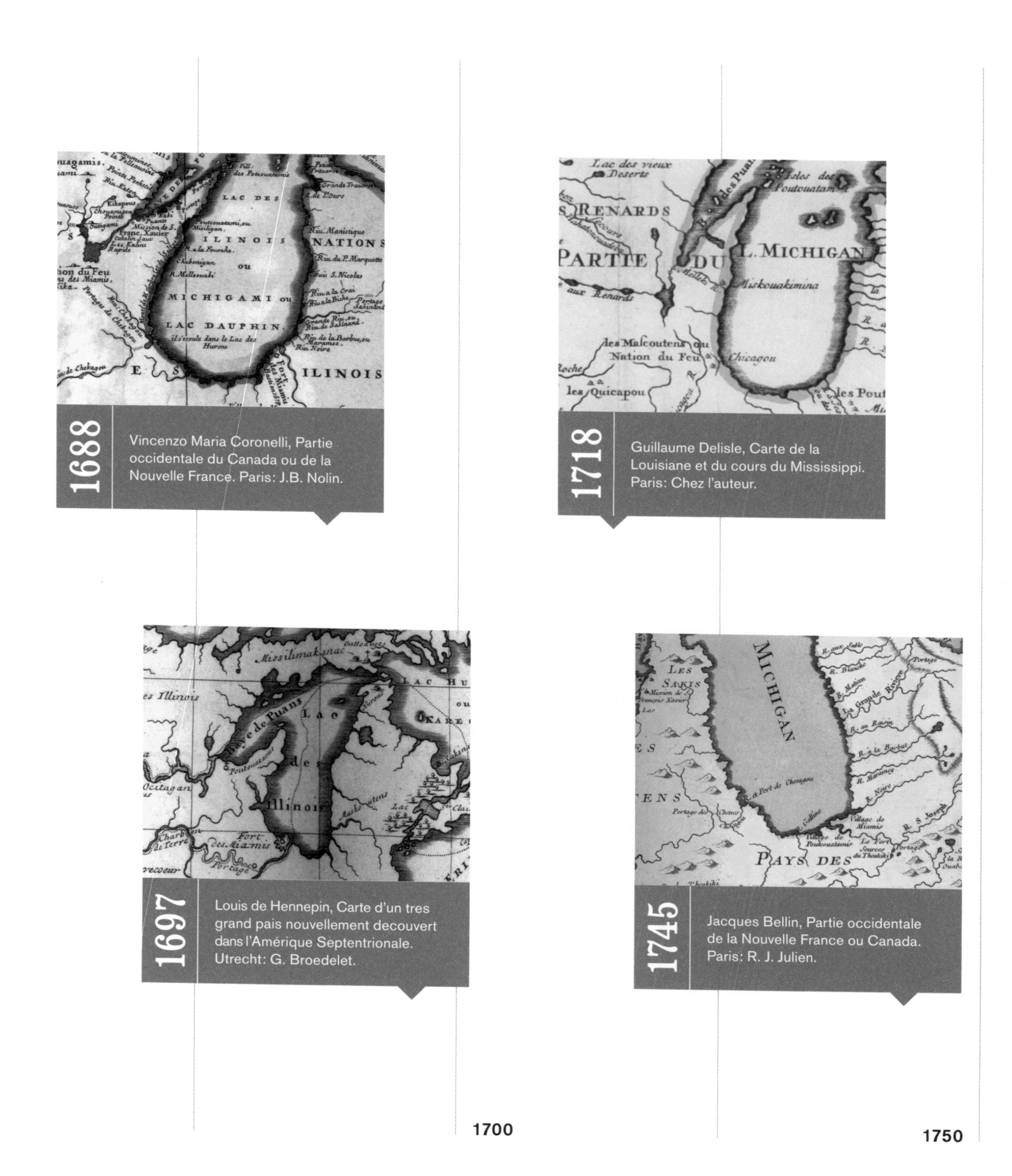

1688 — Vincenzo Maria Coronelli, Partie occidentale du Canada ou de la Nouvelle France. Paris: J.B. Nolin.

1718 — Guillaume Delisle, Carte de la Louisiane et du cours du Mississippi. Paris: Chez l'auteur.

1697 — Louis de Hennepin, Carte d'un tres grand pais nouvellement decouvert dans l'Amérique Septentrionale. Utrecht: G. Broedelet.

1745 — Jacques Bellin, Partie occidentale de la Nouvelle France ou Canada. Paris: R. J. Julien.

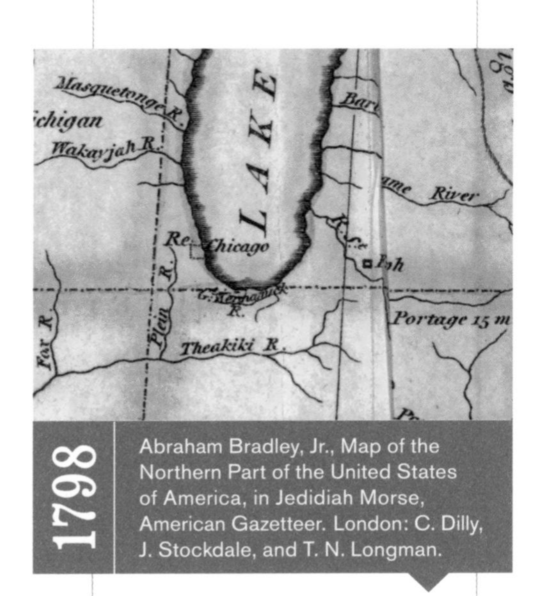

1798 Abraham Bradley, Jr., Map of the Northern Part of the United States of America, in Jedidiah Morse, American Gazetteer. London: C. Dilly, J. Stockdale, and T. N. Longman.

1818 John Melish, United States, in Morris Birkbeck, Letters from Illinois. Philadelphia: M. Carey and Son.

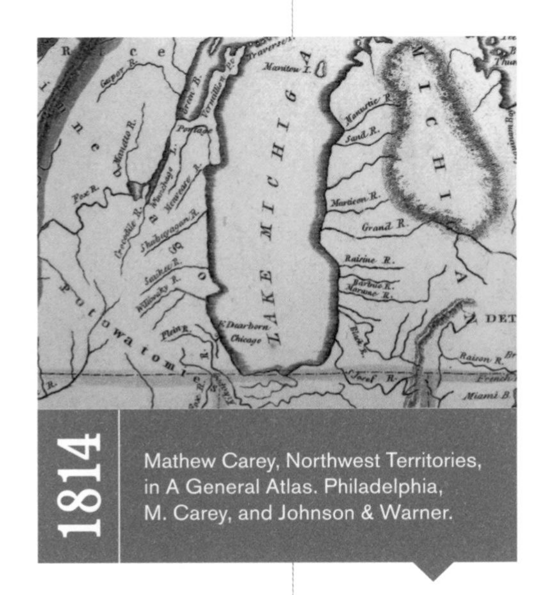

1814 Mathew Carey, Northwest Territories, in A General Atlas. Philadelphia, M. Carey, and Johnson & Warner.

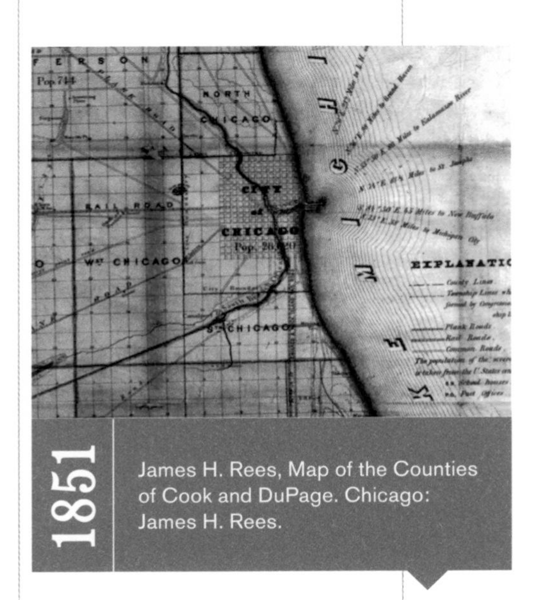

1851 James H. Rees, Map of the Counties of Cook and DuPage. Chicago: James H. Rees.

1800

1850

SECTION TWO: MAPPING TO SERVE THE NEW NATION

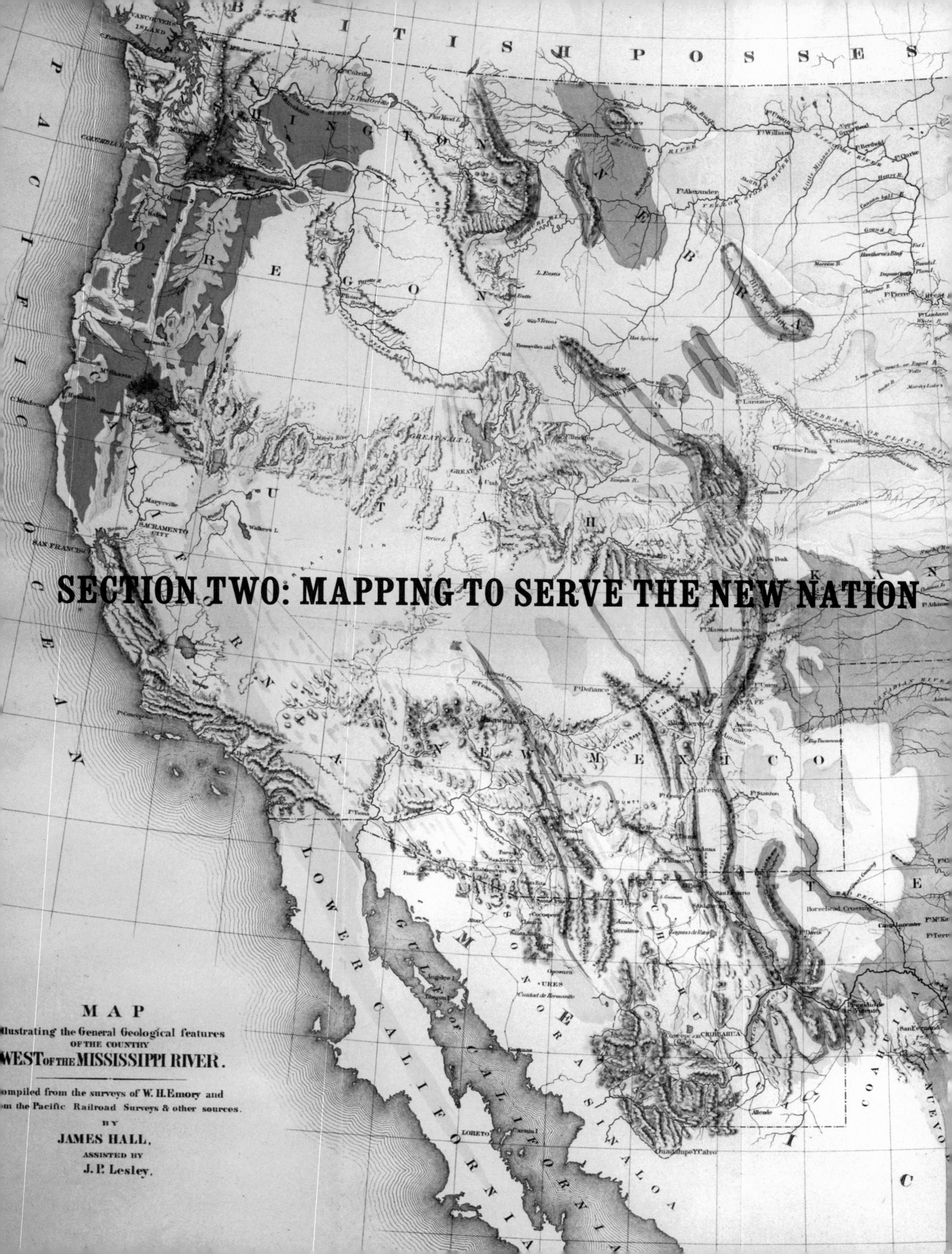

From the earliest days of the United States, maps were vital to the creation of a sense of national identity among citizens, to practical administration, and to the encouragement of economic development. There would be no official government mapping policy until 1925, but ad hoc cartographic programs began to appear shortly after the nation's birth. Over the decades they overlapped in some areas and ignored others. Nevertheless, government maps charted many parts of the West for the first time, leading the way for the commercial cartography and economic development that followed.

Whereas maps of the United States' existing boundaries solidified citizens' sense of geographical and political identity, maps incorporating the Western territories suggested that expansion was equally central to their perceptions of nationhood. To justify the government's appropriation of Western lands, politicians and the press asserted that the extension of the United States across the continent fulfilled a divine mandate. In documenting the North American interior through to the Pacific coast, state-sponsored maps boldly visualized this idea of Manifest Destiny. Maps showing areas west of the nation's official boundaries graphically emphasized the continuity between the states and territories, showing the two as parts of a larger whole. They depicted the region as capable of being traversed, quantified, and understood—and thus ready to be conquered, inhabited, and exploited.

The Age of National Exploration

Before the United States government began systematic exploration, maps of the American West were largely conjectural. Eager to learn more about the trans-Mississippi region, Thomas Jefferson appointed Meriwether Lewis and William Clark to cross the continent from St. Louis to the Pacific Ocean and back. The summary map that Clark drew after the expedition set a new standard for Western cartography, based on field observations and scientific inferences drawn from topography. Shortly after Jefferson received the reports of Lewis and Clark, Congress authorized him to initiate a scientific survey of the nation's coasts, as a vital aid to commerce. The four decades that followed saw few interior expeditions sponsored by the federal government, although the mountain men who continued to explore the region in search of furs supplied much information and advice to politicians and officials.

The government became more directly engaged as Western exploration entered a new phase in the 1840s. The mountain men had fortified the region's reputation as a land of romantic adventure and economic promise, stimulating a renewed interest in territorial expansion. Growing desire to develop trade with Asia and to secure the nation's western borders also contributed to this revival. To facilitate the "Great Reconnaissance" of the West, the Army reorganized the Corps of Topographical Engineers in 1838, creating a unit of officers trained in scientific exploration and mapmaking. Politicians soon dispatched the Corps to chart new paths to the Pacific, build wagon roads, and collect data to settle boundary disputes with England in Oregon and with Mexico in the Southwest. The Coast Survey began mapping the Western seaboard in 1850.

2.1

The Map That Made the West American

CREATORS: William Clark (cartographer), Samuel Lewis (engraver)

TITLE: *A Map of Part of the Continent of North America*, c. 1810. Pen and ink on paper

SIZE: 32 x 52 inches

LOCATION: Collection of Western Americana, Beinecke Rare Book and Manuscript Library, Yale University, New Haven, CT

William Clark
Charles Willson Peale, *William Clark by Charles Willson Peale, from Life* (1807–1808)
CREDIT: Independence National Historical Park

Meriwether Lewis
Charles Willson Peale, *Meriwether Lewis by Charles Willson Peale, from Life* (1807–1808)
CREDIT: Independence National Historical Park

The storied Lewis and Clark expedition was a geographical and scientific endeavor prompted by economic motives. President Thomas Jefferson conceived the project in 1802, outlining three major goals: opening an easy route to the Pacific to facilitate trade with Asia; finding ways to develop the fur trade; and locating farmland. The endeavor gained urgency after the 1803 Louisiana Purchase added some 828,000 square miles of western territory to the United States.

To carry out the plan, Jefferson appointed Meriwether Lewis (1774–1809) and William Clark (1770–1838), who recruited a 40-member Corps of Discovery. Between May 1804 and September 1806, the party traversed more than 8,000 miles. Along the way, Lewis made observations about the plants, animals, and native people they encountered in his journal, while Clark served as the cartographer. In the evening, Clark would take out a quill pen and rag paper, mix some powdered ink, and record on a field map the geographical observations, compass bearings, and knowledge gleaned from Native Americans that day. Over the course of the expedition, Clark drew nearly 200 maps.

After the party returned, Clark synthesized his data on a "master map," completed in 1810. He outlined the route of the Corps, adding information about peripheral areas gathered from Indian informants and from the findings of other explorers. Clark sent the manuscript map to Philadelphia, where Nicholas Biddle, assisted by Corpsman George Shannon, was editing the expedition's journals for publication. Samuel Lewis engraved the map for inclusion with the explorers' report, published in 1814. In addition to serving as the primary visual record of the expedition, the map became the model for maps of the West for the next half century.

As Clark's map demonstrated, the region was rich in fur-bearing animals, fish, and game, but also included vast areas of dry plains ill-suited to agriculture. The hoped-for river route to the coast did not exist, and the towering Rocky Mountains presented a formidable barrier to overland passage. Despite these obstacles, Clark's great map provided tangible evidence of the reach and ambition of the new nation. It staked a claim to the Oregon Territory, inspired people to think of a United States stretching from coast to coast, and cemented a relationship between the federal government and Western mapping that would last a century.

References: Ambrose 1996; Cohen 2002; Moulton 1983; Moulton 2003

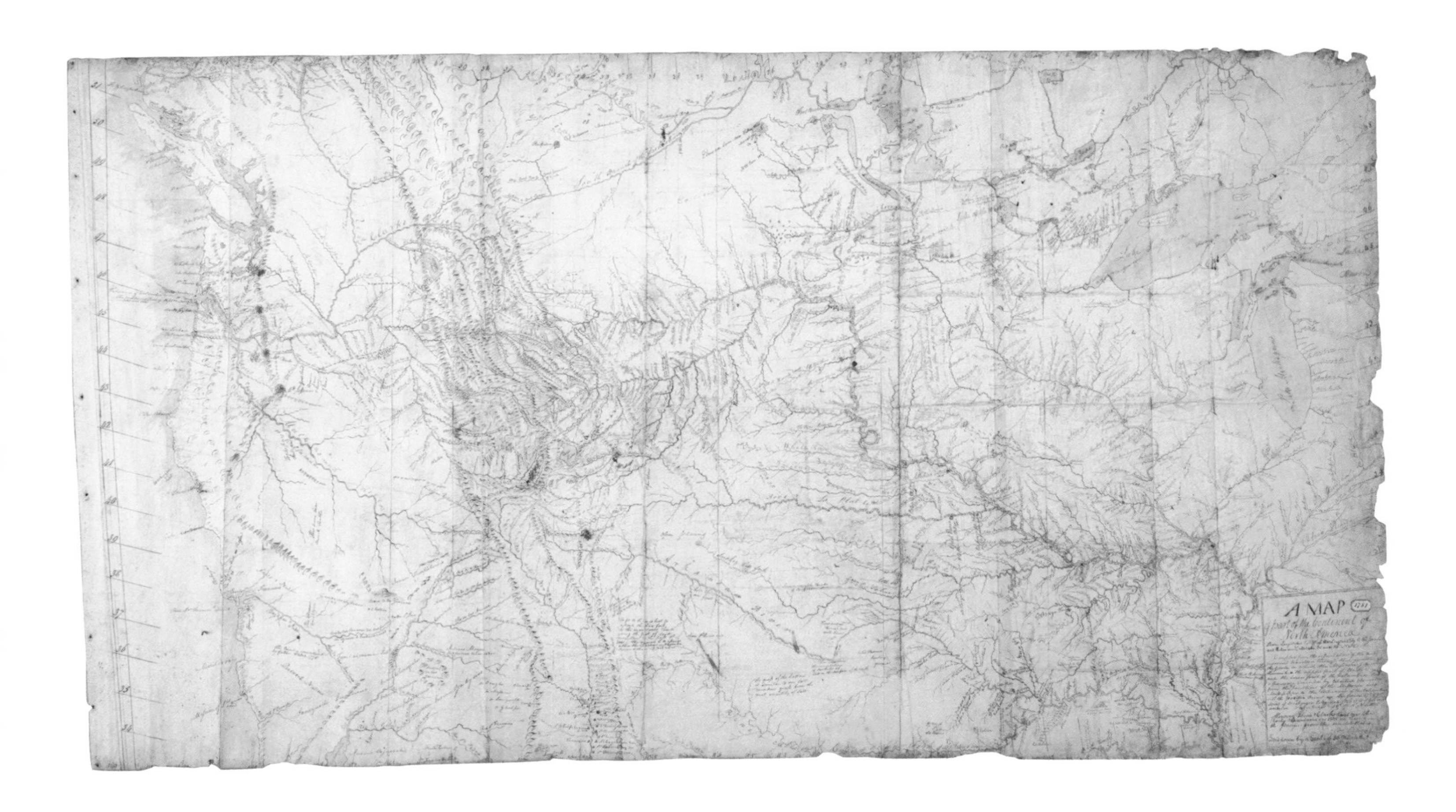

Explorers, Mapmakers, Heroes

CREATORS: Charles Preuss (cartographer), annotated by George Gibbs with information from Jedediah Smith, E. Weber & Co. (lithographers)

TITLE: *Map of an Exploring Expedition to the Rocky Mountains in the Year 1842 Oregon & North California in the Year 1843–44 by Brevet Capt. J. C. Frémont of the Corps of Topographical Engineers Under the Orders of Col. J. J. Abert, Chief of the Topographical Bureau* (Baltimore, MD: E. Weber & Co., 1845, annotated 1849–1851), from *Report of the Exploring Expedition ...*, in Senate Executive Document 174 (Washington, DC: Gales and Seaton, 1845)

SIZE: 31.5 x 51 inches

LOCATION: American Geographical Society Collection, University of Wisconsin-Milwaukee

Charles Preuss
Century vol.19 (1891): 759.
CREDIT: Newberry A5.19

Jedediah Smith
Jedediah Smith Society Collection mss 18, 4.1.10
CREDIT: Holt-Atherton Special Collections, University of the Pacific Library

This map, published with John C. Frémont's report of his 1842 and 1843–44 journeys, illustrates the evolving motives and uses that shaped government exploration and mapmaking. Missouri Senator Thomas Hart Benton instigated the expeditions to gather information to support American acquisition of Oregon. Senator Benton secured the appointment of his son-in-law Frémont (1813–1890) to the Corps of Topographical Engineers to lead the reconnaissance. The report and map, drawn by German-born cartographer Charles Preuss (1803–1854), also served the broader purpose of rallying public interest in Westward expansion. Together they offered the first comprehensive, scientific view of the region. The charismatic Frémont, nicknamed "the Pathfinder," helped create the popular image of the explorer as romantic hero. His lively report, ghostwritten by his wife Jessie, became a bestseller, while knockoffs of the map soon appeared in countless emigrant guides to the West.

John Charles Frémont
Evart Duyckinck, *National Portrait Gallery of Eminent Americans* vol. 2: opposite p. 329.
CREDIT: Newberry E 483.24

George Gibbs (1815–1873), a Harvard-educated New Yorker, personalized his copy of the map with extensive inscriptions. In 1849 Gibbs succumbed to the lure of the West, joining an army regiment bound for Oregon. He traced the route of his regiment's march and elaborated its topographical details with extensive pencil notes on the map. For his annotations, Gibbs drew on a now-lost manuscript map made in 1831 by Jedediah Smith (1799–1831), one of the early West's greatest explorers. Because Smith's work was never published, Gibbs's inscribed map is the most complete surviving record of the legendary mountain man's wealth of Western knowledge, gathered in the course of his fur-trapping expeditions in the 1820s. Gibbs recorded Smith's remarkable crossing of the Sierra Nevada Mountains and the Great Basin (in what is now Nevada and Utah) in 1826–1827, years before the area was documented on official government maps. Smith's understanding of the West's interior was not captured on published maps until the mid-1850s.

References: Goetzmann 1959; Goetzmann 1966; Morgan 1953; Morgan and Wheat 1954; Wheat 1957–1963

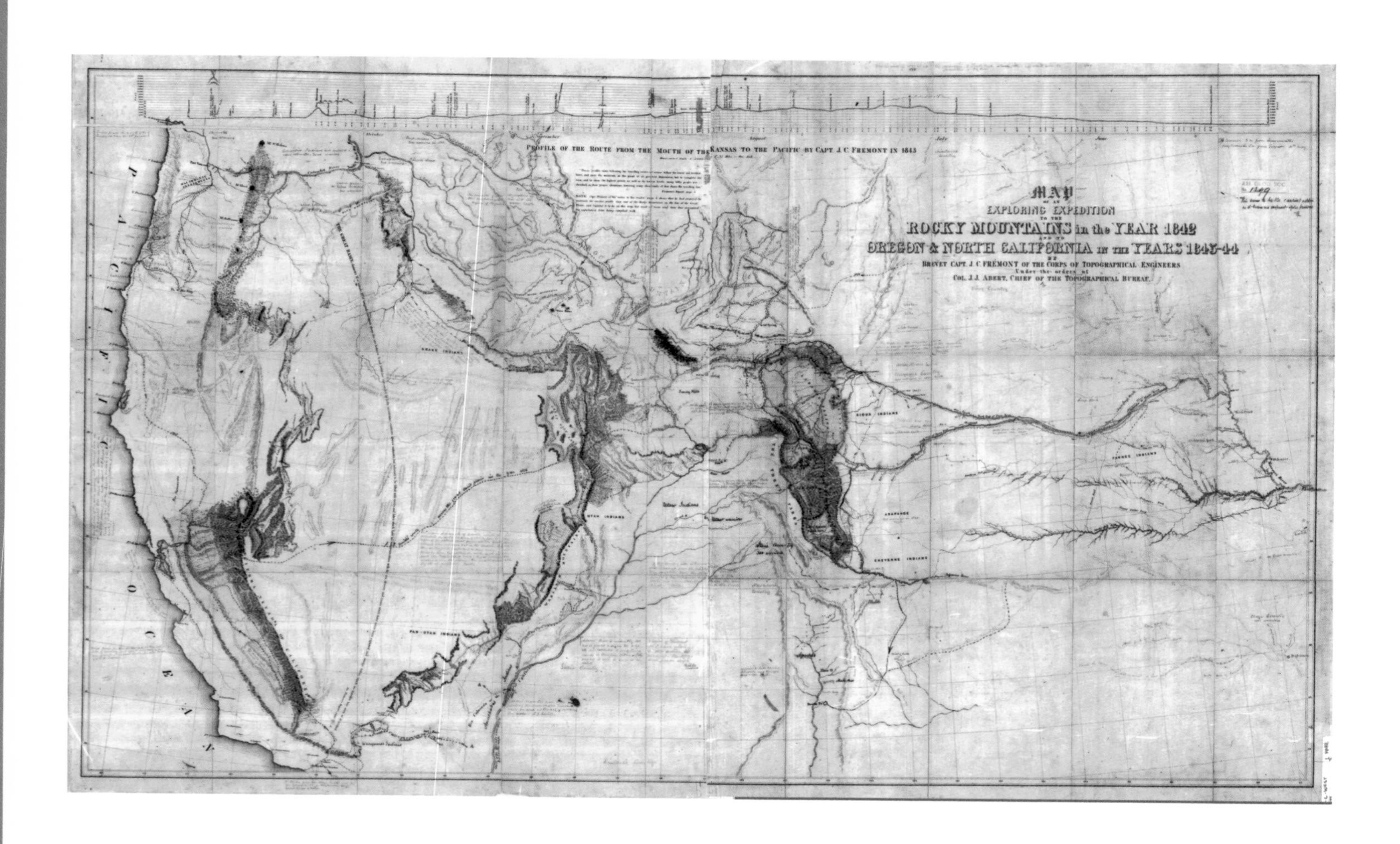

2.3

Early Geologists in the West

CREATORS: James Hall, assisted by J. Peter Lesley (cartographers); Thomas Jekyll (draftsman); Sarony, Major & Knapp (lithographers)

TITLE: *Map Illustrating the General Geological Features of the Country West of the Mississippi River* (New York: Sarony, Major & Knapp, 1857), from William Hemsley Emory, *Report on the United States and Mexican Boundary Survey*, in U.S. 34th Congress, 1st session, 1855–1856. Senate Executive Document 108 (Washington, DC: Nicholson, 1857)

SIZE: 20.8 x 23.6 inches

LOCATION: Newberry Map 4F G 4051 C5 1857 H3, copy 1

James Hall
Harper's Weekly 28 (September 27, 1884): 630.
CREDIT: Newberry A5 391

J. Peter Lesley
Harper's Weekly 28 (September 27, 1884): 630.
CREDIT: Newberry A5 391

The government surveys also provided data for more specialized, scientific maps. James Hall and J. Peter Lesley compiled information from several projects, including the Mexican Boundary Survey of 1848–55 and Pacific railroad surveys of 1853–55, to produce the earliest colored geological map of the West published by the government. The map accompanied William H. Emory's official *Report on the United States and Mexican Boundary Survey*, as well as the report of the Pacific railroad surveys. Hall and Lesley adapted the summary map of the trans-Mississippi West produced by Emory, the head of the boundary survey's topographical unit. Their map was folded into the second part of the report, a scientific study titled *Palaeontology and Geology of the Boundary*.

A prominent geologist and state paleontologist for New York, Hall (1811–1898) served as a consultant to the Mexican Boundary Survey. Lesley (1819–1903) worked for the state geological survey of Pennsylvania before becoming Secretary of the Iron Association in 1855 and the American Philosophical Society in 1858.

The map illustrates the extent—and limits—of geologic knowledge of the West at the time. Hall and Lesley identified the mountain ranges as metamorphic formations (shown in pink), most often flanked by bands of Cretaceous and Carboniferous rocks (light green and blue). They also indicated large areas of igneous rocks (colored orange) in the Pacific Northwest. But they left the entire Great Basin region and the western part of the Oregon territory, on either side of the Snake River, almost entirely blank, as no real scientific surveys had yet been conducted of these areas. The latitudes under consideration for the transcontinental rail route are elaborated in detail, reflecting the abundance of data gathered by the Pacific railroad surveys.

Moreover, the information they included reflected the intellectual concerns of Eastern scientists rather than the practical demands of people on the frontier. Hall and Lesley focused on stratigraphy at the expense of utilitarian data, such as the location of coal, salt, and precious metals. The next generation of scientists, who produced maps of the West after the Civil War, would prove more attentive to the needs of settlers and developers on the ground.

References: Goetzmann 1966; Merrill 1927

2.4

Art and Science at the Grand Canyon

CREATOR: William Henry Holmes

TITLE: *The Grand Cañon at the Foot of the Toroweap—Looking East,* in Clarence E. Dutton, *Atlas to Accompany the Monograph on the Tertiary History of the Grand Canyon District* (New York: Julius Bien & Co., 1882)

SIZE: 20.5 x 36 inches

LOCATION: Newberry Ayer 109.9 G4 D9 1882 atlas [not on display]

CREDIT: Gift of Edward E. Ayer

William Henry Holmes

William Henry Holmes by Rock Formations

CREDIT: National Anthropological Archives, Smithsonian Institution (NAA INV. 02859400)

With his celebrated panoramas, William Henry Holmes (1846–1933) brought the West's breathtaking scenery to audiences across the country. Directing his artistic skills to scientific ends, Holmes drew topographical and geological sketches for the surveys led by Ferdinand Vandeveer Hayden and John Wesley Powell. In 1880 he traveled to the Grand Canyon to work with Clarence E. Dutton (1841–1912), whom Powell had tapped to write a geological history of the region. An elephant folio atlas, featuring chromolithographs of Holmes' panoramic watercolors, accompanied the limited edition.

Together, Holmes's pictures and Dutton's words created a portrait of the canyon as poetic as it was precise. Leading readers through the Grand Canyon district, Dutton explained how the rock from the Tertiary period and several other geologic ages had been weathered away by wind and water, sun and frost. By capturing hundreds of miles of terrain in stratigraphic detail, Holmes's paintings visualized the landscape better than would maps or photographs. The most dramatic image was the three-part panorama viewed from a spot Dutton called Point Sublime. From here the canyon loomed, in Dutton's words, as "by far the most sublime of all earthly spectacles."

References: Goetzmann 1966; Goetzmann 1979; Goetzmann 1986; Swinton 1937; Worster 2001

Dispossessing Native Peoples

Because the federal government's Indian policy in the West was at its core geographical, maps proved well suited to schematize it. The policy focused on ejecting Native Americans from areas sought by whites, relegating them to increasingly smaller and less desirable parcels. In 1830 Congress passed the Indian Removal Act, authorizing the President to negotiate with all eastern nations in order to relocate them to land west of the Mississippi. When the Cherokee resisted, U.S. troops forced them to march more than a thousand miles, a route the Indians called the "Trail on Which We Cried." One in eight Cherokees died on the trail or soon after arriving in Indian Territory. Over the course of the nineteenth century, Indians from various parts of the country were displaced from their homelands.

Although removal promised independence in the West, the government gradually eroded the Indians' sovereignty along with their land holdings. Federal policy shifted from conceptualizing tribes as small nations to regarding Indians as individual wards of the United States. Officials created reservations around the West as a way to ensure racial segregation and discourage conflicts between Indians and whites. Politicians and reformers alike embraced the contradictory position that the forced isolation of Native Americans would prepare them for integration into white society.

Policymakers saw the substitution of individual property ownership for communal land holding as one of the linchpins to assimilating the Indians, along with the government-supervised education of children and the suppression of traditional patterns of family and community life. They regarded the forced adoption of Euro-American practices of landownership as integral to the process of transforming Native Americans into farmers and ranchers. To advance this policy, the 1887 Dawes Act provided for the division of reservation land among native peoples. Maintaining that Indian populations were declining and many reservations contained more land than the Indians needed—especially after they gave up hunting for farming—the law also allowed the sale of surplus land to whites. The often-reckless implementation of the Act sped the process of dispossession. Indian property holdings dropped from 155,632,312 acres in 1881 to 77,865,373 acres by 1900.

Intimidating the Navajos in New Mexico

CREATORS: Edward M. Kern (cartographer), J. J. Abert, (Chief, U. S. Army Crops of Topographical Engineers), J. G. Shoemaker (engraver), P. S. Duval (printer)

TITLE: *Map of the Route Pursued in 1849 by the U. S. Troops, under the Command of Bvt. Lieut. Col. Jno. M. Washington, Governor of New Mexico, in an Expedition against the Navajo Indians*, from J. H. Simpson, *Journal of a Military Reconnaissance, from Santa Fé, New Mexico, to the Navajo Country: Made with the Troops under Command of Brevet Lieutenant Colonel John M. Washington, Chief of Ninth Military Department, and Governor of New Mexico, in 1849* (Philadelphia: Lippincott, Grambo, and Co. 1852)

SIZE: 20 x 27.6 inches

LOCATION: Newberry Ayer map 4F G4321 E1 1849 S5

CREDIT: Gift of Edward E. Ayer

The Navajos met little interference from early Spanish colonizers, who focused their attentions on lands further south. But after Mexico ceded the territory in 1848, the American government sought to assert its new authority. Lieutenant Colonel John Washington, territorial governor of New Mexico, organized his 1849 expedition chiefly to intimidate the Indians who had inhabited the region for centuries—to halt Navajo raids on the towns along the Rio Grande and pave the way for future settlement.

Washington appointed James H. Simpson (1813–1883), an officer in the Army Corps of Topographical Engineers, to accompany the expedition. Simpson in turn hired two civilians, Edward M. Kern (1823–1863) and his brother Richard H. Kern (1821–1853), to produce maps and views of the territory. Edward Kern's map, published with Simpson's journal of the expedition, traced the party's route in red. Kern represented in detail the topography of the Chaco Canyon and Canyon de Chelly, two Navajo strongholds. Washington's troops defeated the Indians at both locations, permanently defusing their power.

After the military mission was accomplished, Simpson and the Kerns were freed for scientific explorations. They examined some of the most important prehistoric sites in the southwest, where Simpson penned descriptions in his journal, Edward Kern drew maps, and Richard Kern made sketches of native peoples, architecture, and geological formations. The trio closely studied Pueblo Bonito, the garrisoned capital of Chaco Canyon, which stretched across almost two acres and had become a command center when Anasazi culture flourished between 1100 and 1300. At Inscription Rock near Zuñi, the party made copies of the voluminous inscriptions in Spanish, Latin, and Indian "hieroglyphics."

Simpson's orders also directed him to explore the Old Spanish Trail from Santa Fe to Los Angeles with an eye toward finding a route for the transcontinental railroad. Adding information from mountain man Richard Campbell to his own findings, Simpson identified a viable road due west from the Rio Grande through Zuñi. The topographical rendering on Kern's *Map of the Route Pursued in 1849 by the U. S. Troops*, along with the table of distances and notes about the availability of water and pasture, summarized information useful to railroad planners. More generally, the map offered the first accurate portrayal of New Mexico's geography, and foreshadowed subsequent encounters between the Army and the Indians, when troops chased natives not merely to frighten them but to corral and contain them within reservations.

References: Goetzman 1966; Hine and Faragher 2000; Simpson 1964

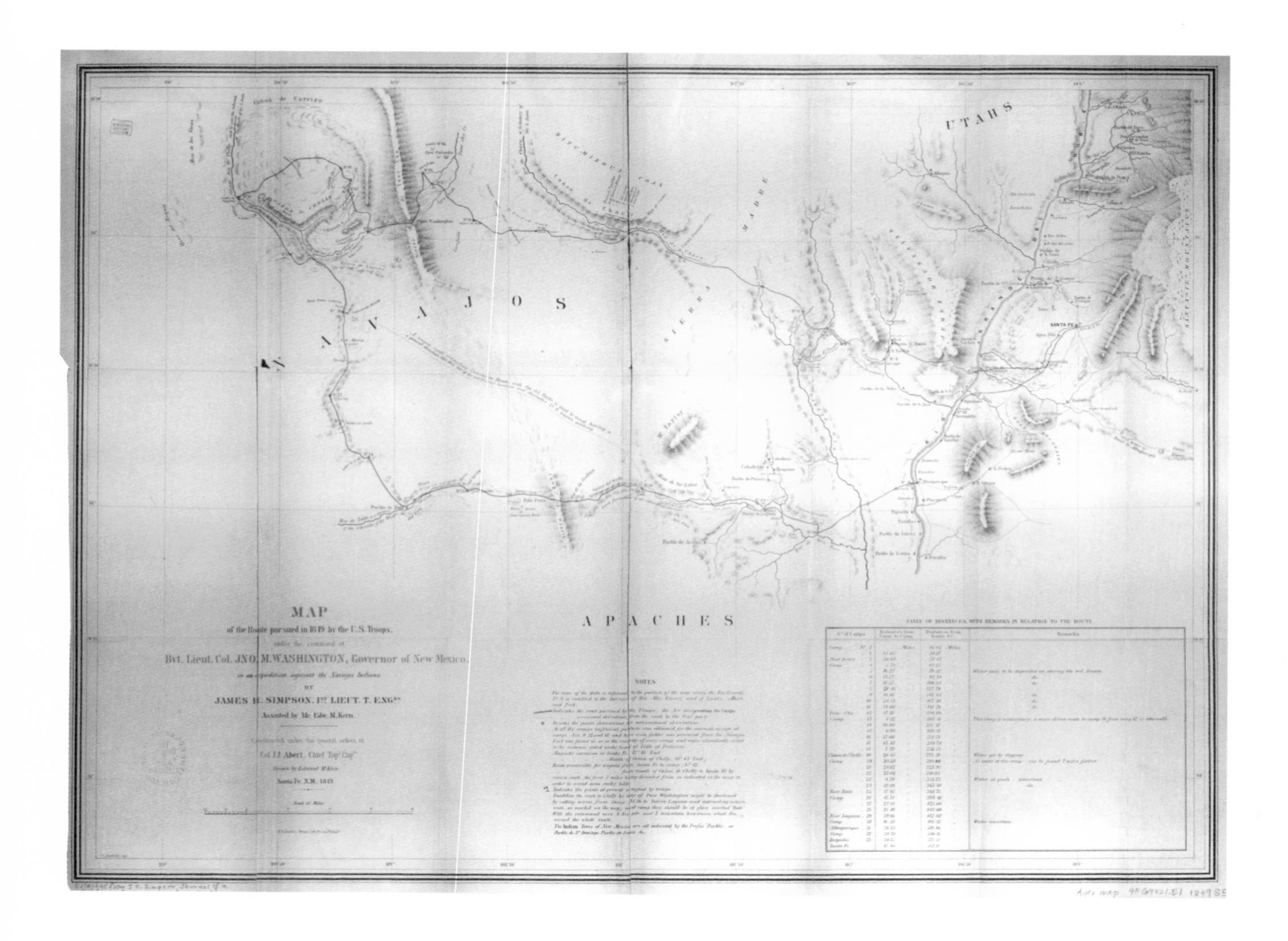

2.6

Mapping the Nez Percé Tragedy

CREATORS: Oliver O. Howard (commander), Robert H. Fletcher (draftsman)

TITLE: *Map of the Nez Perce Indian Campaign*, from U.S. House 45th Congress, 2nd session, *Report of the Secretary of War, 1877.* (Washington, DC: Government Printing Office, 1878)

SIZE: 24 x 47.5 inches

LOCATION: Newberry Map 6F G4241 55 1878 F5

Oliver Otis Howard

Century 32 (September 1886): 752.

CREDIT: Newberry A5 19

Beginning in the 1840s, the flood of white emigrants into the Oregon country of the Pacific Northwest prompted conflicts with the many small groups of Native Americans established there. To clear the most desirable land for white settlers, federal officials negotiated a series of treaties with the Indians, creating an extensive system of reservations in the Oregon and Washington territories. Among the Nez Percé, a group of independent bands united by language, some groups ceded their land in 1855 and removed to a reservation; others refused. In 1863, after miners discovered gold on Nez Percé land, the government reduced the acreage of the reservation. White settlers continued to encroach on the land held by bands whose leaders had not signed the treaty, increasing tensions not only between Indians and whites, but also between the bands who accepted the treaty and those who did not.

The conflict reached a dramatic climax in the Nez Percé War of 1877, the course of which is outlined in the *Map of the Nez Perce Indian Campaign.* In May, General Oliver O. Howard (1830–1909) demanded that the non-treaty bands move to the newly reduced reservation. After the chiefs reluctantly agreed, four angry young Nez Percé men killed four white ranchers. In the ensuing panic, most of the anti–treaty bands fled to the mountains led by Chief Joseph (1840–1904), Looking Glass, White Bird, Toohoolhoolzote, and Ollokot. Howard pursued them with 500 soldiers and army volunteers.

The map published with the annual report of the Secretary of War traces the route of the military as they followed the 1,300-mile, three-month flight of the Nez Percé. The draftsman, Army Lieutenant Robert Fletcher, was the chief signal officer for the Department of Columbia, under Howard's command. His hand lettering and anecdotal details, such as the flags identifying forts and Indian agencies, liken the map to folk art. Fletcher illustrated the major battle sites in the vignettes along the lower border. At White Bird Canyon and along the Clearwater River in Idaho, at Big Hole in Montana, and at other points, the Indians fought skillfully against larger, better-equipped forces.

After their old allies, the Crows, refused to assist them, the Nez Percé decided to escape to Canada. When they reached Bear Paw Mountain, forty miles south of the border, General Nelson A. Miles cornered the exhausted Indians. A few hundred were able to slip through to Canada, where they joined the Hunkpapa Sioux Chief Sitting Bull (1834?–1890). On October 5, 1877, Chief Joseph formally surrendered for the remaining four hundred Nez Percé, making a speech that would become famous: "It is cold and we have no blankets. The little children are freezing to death. . . . I am tired. My heart is sick and sad. From where the sun now stands, I will fight no more forever."

References: Hoxie 1996; Lavender 1992; Walker 1998; White 1991a

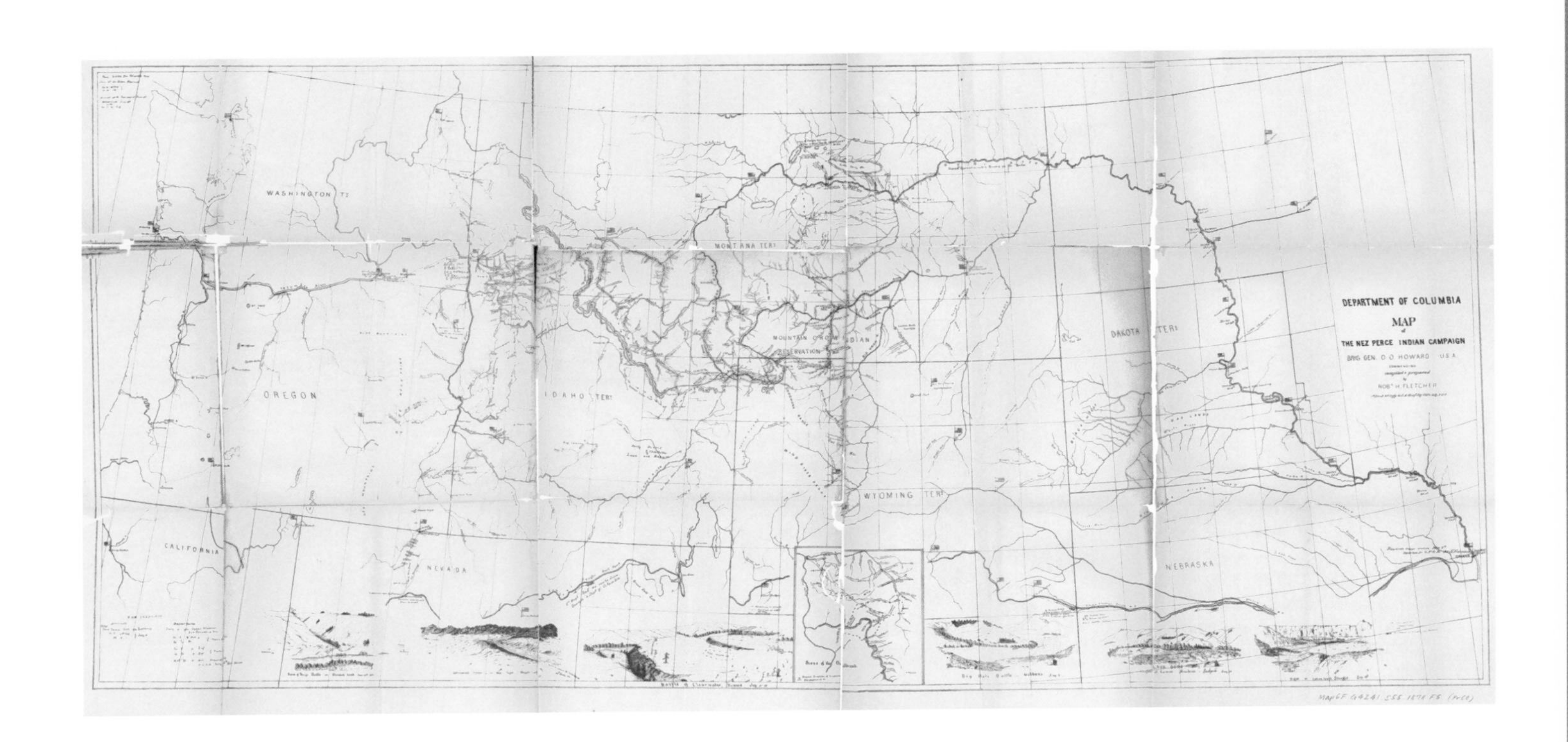

2.7

Parceling Indian Territory

CREATORS: C. A. Maxwell (compiler), John Olberg (draftsman)

TITLE: *Indian Territory Compiled under the Direction of the Hon. John H. Oberly, Commissioner of Indian Affairs*, from *Letter from the Secretary of the Interior, Transmitting, in Response to Senate Resolution of March 10, 1890, the Compilation concerning the Legal Status of the Indians in Indian Territory*, in U.S. 51st Congress, 1st session, 1889–1890. Senate Executive Document 78 (Washington, DC: Government Printing Office, 1890)

SIZE: 23.6 x 32 inches

LOCATION: Newberry Map 6F G4021 E1 1889 U5

In 1889, as the United States prepared to open the western half of the land set aside as Indian Territory to white settlers, the Office of Indian Affairs published a map summarizing the history and current status of land assignments within the territory. The government published the map with an 1890 Senate document that explained the provisions of treaties made between the government and the Indians over the course of the nineteenth century.

The mapmakers used boldly-colored outlines to demarcate the lands allocated to each Native American group, whose position reflected their original locations. Thus the Five Civilized Tribes (Cherokee, Choctaw, Creek, Chickasaw, and Seminole) dominate the southeastern section of the territory; the Midwestern Iowa, Sac, and Fox occupy the north central portion; the Cheyenne reside further west; the Comanche in the southwest; and so forth. The titles and dates of the treaties by which the Indians agreed to exchange their homelands for acreage in the territory, along with references to the published treaties, are printed within each parcel. This information, along with the colored outlines, was probably overprinted on a pre-existing general topographic map of the territory.

The map includes several signs of the intrusion of white culture, such as towns, military reservations, railroads, and land divisions. In accordance with the Dawes Act, most of the territory depicted has been surveyed and divided into townships, in preparation for selling parcels to whites. Over 50,000 homesteaders rushed into the territory on the day it opened—April 22, 1889. Because the Dawes Act exempted the Five Civilized Tribes, land belonging to the Cherokee, Creek, Choctaw, and Seminole remains undivided. (The Chicasaws had already negotiated with the government for the allotment and sale of their land.) In less than a decade, all land in the territory would be parceled for sale.

References: Hoxie 1996; Morris, Goins, and MacReynolds 1976; United States Senate 1890; Washburn 1988; White 1991a

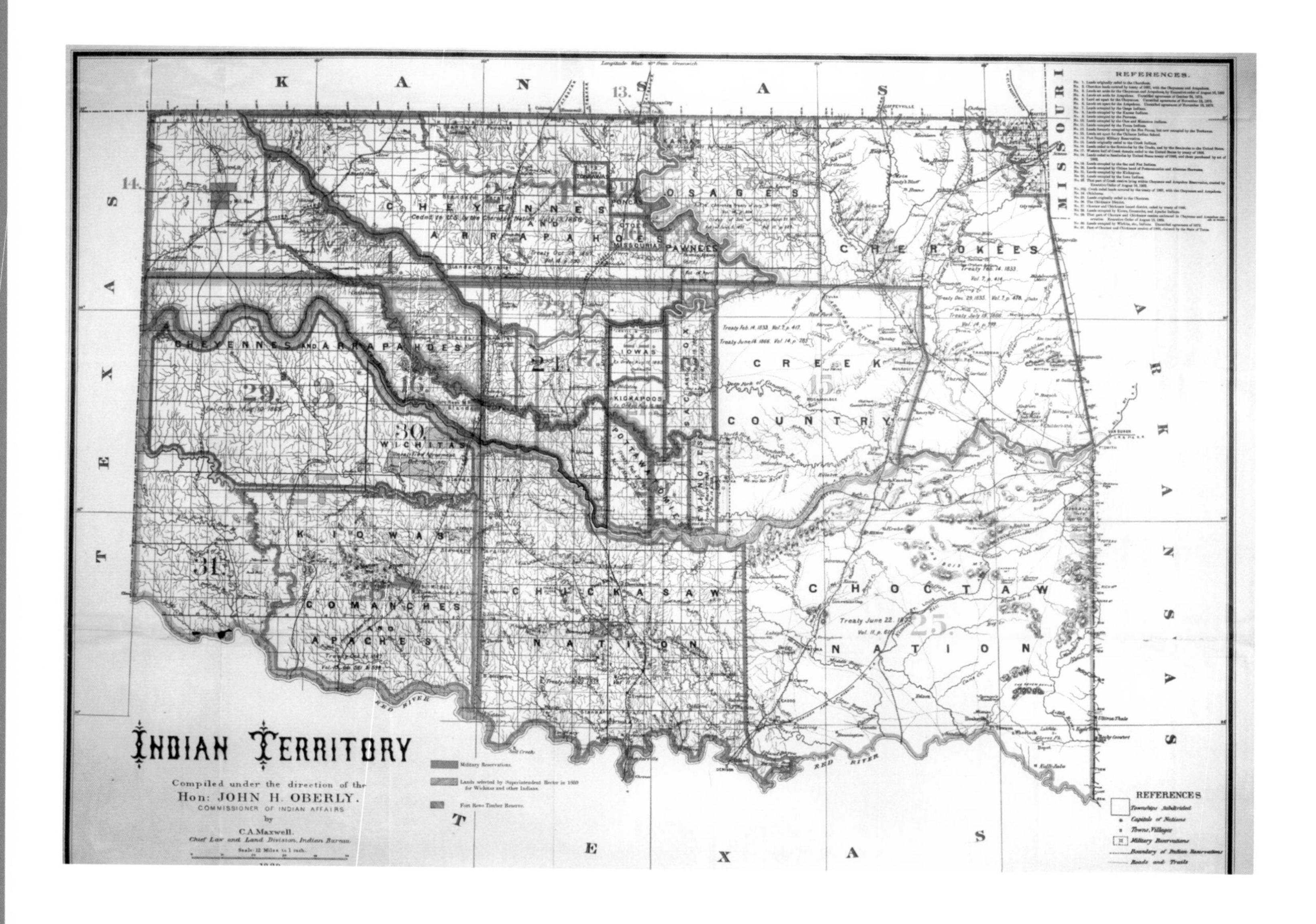

Surveying Western Resources

After the Civil War, the government launched a multi-pronged program of economic expansion in the West. Congress directed substantial federal resources to building a privately-owned transcontinental railroad system, and commissioned related inventories of Western minerals, timber, and arable land. Absent any national mapping agency or policy, the government commissioned four scientific surveys, whose coverage sometimes overlapped. Each survey scrutinized topographical and geological features with an eye toward profitable exploitation.

These official projects generated a staggering volume of maps and data about the West. By providing useful information without charge, the government indirectly subsidized the corporate development of the region. As a result, the maps illustrate the early transformation of the Western public domain into valuable private property in the form of transportation networks, mining claims, lumber districts, farms, and ranches.

Military Mapping in Service to the Railroad

CREATORS: Grenville Mellen Dodge (in charge), George T. Robinson (engineer), T. H. Williams (draftsman), Christopher Columbus Augur (contributor)

TITLE: *Map of the Military District, Kansas and the Territories*, 1866. Red, brown, black, and blue ink and ink washes on paper

SIZE: 47.72 x 67.28 inches

LOCATION: Newberry vault MS Map 6F G4050 1866 D6

CREDIT: Gift of Art and Jan Holzheimer

Christopher Columbus Augur
Portrait Monthly, vol.1, no. 2 (August 1863): 23.
CREDIT: Newberry folio E 4834.7

Grenville Mellen Dodge
Annals of Iowa, vol.4, no. 3 (July 1866): opposite 673.
CREDIT: Newberry F 912.01

The manuscript *Map of the Military District* opens a window onto the Army's post-Civil War Western operations, the planning of the transcontinental railroad, and the creation of Western myths. General Grenville Mellen Dodge (1831–1916) ordered the map as commander of the Western Plains territory. Dodge had surveyed railroad routes in the West before joining the Union Army during the Civil War. After the war, railroad barons who knew he was sympathetic to their interests lobbied for his appointment to the Plains post. Implicitly, Dodge was charged with driving Indians away from the projected route of the transcontinental railroad.

The map's careful enumeration of topographic detail, military forts, mining camps, army expedition routes, wagon roads, and trails assured its usefulness to Dodge's task. To produce the map, supervising engineer Major George T. Robinson and draftsman T. H. Williams synthesized and updated information from other army maps. General Christopher Columbus Augur (1821–1898), commander of the Army's Department of the Platte, also contributed data. The division of the map into 28 conveniently-sized panels, backed with cloth, was typical of the era's army field maps.

Dodge served as a consultant to the Union Pacific Railroad while on active duty and resigned his commission in May 1866 to work full-time as the railroad's chief engineer. It seems likely that he took the map with him for use in his new job. The manuscript includes several revisions made after Dodge left the army, most notably in southern Wyoming.

The map offers numerous clues to the process of planning the rail route. The stretch from Omaha to North Platte, Nebraska, is shown as a black line crossed with perpendicular bars, indicating that the track had already been laid. The next section, from North Platte to the vicinity of Cheyenne pass, is indicated with an uncrossed line, probably signifying a route that had been determined but not yet built. Further west, the notations become more tentative. The areas that the railroad ultimately traversed are marked by lines and dashed lines drawn in pencil, clouded by numerous erasures. These sections lack the carefully-inked hachuring of the surrounding panels. The cartographer probably copied the topography of the more finished-looking panels from other maps, as these sections document areas beyond the proposed railroad route.

In later years, Dodge used the map to help fashion his image as a Western legend. In his memoirs, Dodge aggrandized his role in building the Union Pacific, falsely claiming that his discovery of a mountain pass and meetings with President Lincoln were instrumental in determining the route. He had similar illusions about the map's influence, asserting that the government "had to print a very large number" because "every officer of the Government wanted them." But the only known copies of the map are four photographic replicas that reproduce an earlier state on a smaller scale. Dodge's tendency to bend the truth to his own purposes is also reflected in the omission of any mention of Native Americans. Through this erasure, the map implies that Dodge had thoroughly removed this challenging human obstacle from the railroad's path.

References: Farnham 1965; Goetzmann 1959; Goetzmann 1966; Hirshson 1967; Parke-Bernet Galleries 1969; Thurman 1989

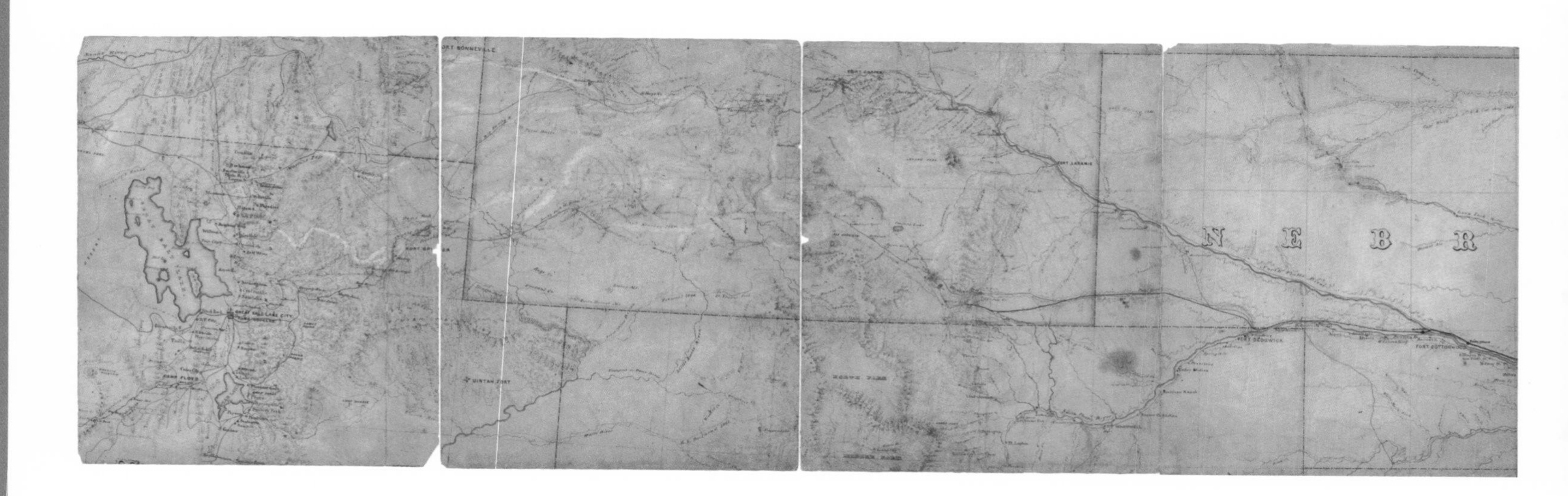

2.9

Inventorying Desert Regions

CREATORS: Department of the Interior, U. S. Geographical and Geological Survey of the Rocky Mountain Region, John Wesley Powell (in charge), Charles Mahon, John Henry Renshawe, Walter Hayden Graves, and Henry Lindenkohl (compilers, draftsmen)

TITLE: *Map of Utah Territory, Representing the Extent of the Irrigable, Timber, and Pasture Lands,* from J[ohn] W[esley] Powell, *Report on the Lands of the Arid Region of the United States...* in U.S. 45th Congress, 2nd session, 1877–1878. House Executive Document 73 (Washington, DC: Government Printing Office, 1878)

SIZE: 36 x 28.5 inches

LOCATION: Newberry Map 6F 4341 G434 1878 U5

John Wesley Powell

Critic 41 (November 1902): 399.

CREDIT: Newberry A5 252

John Wesley Powell led one of the four major surveys of the West after the Civil War, focusing on the Colorado River and the surrounding high plateaus of the Utah and Arizona territories. After a much-heralded adventure down the Colorado River in 1869, which included the first successful navigation of the river through the Grand Canyon, his team took seven more summer-fall seasons to complete the detailed topographical mapping and collection of natural history specimens from the area. Powell published their findings in his 1878 *Report on the Lands of the Arid Region of the United States,* including the map of Utah Territory in a pocket inside the book's back cover.

Incorporating data from the three other government surveys along with the information gathered by Powell's team, the map's hachures traced Utah's rugged topography. Splashes of green identified land suitable for farming and pasturing, blue marked areas of standing timber, and tan indicated forests destroyed by fire. The map also summarized the existing networks of railroads, wagon roads, trails, and telegraph lines.

For Powell, the survey's practical implications lay in its information about the region's agricultural potential. The map's bold colors dramatized the survey's conclusion that merely 2.8 percent of the entire territory could be irrigated, and that the irrigable lands were irregularly dispersed across the territory. Powell recognized that this made the rectangular system of land division impractical. If one farmer's parcel included all the water for any given location, the surrounding land could not be cultivated. Adapting the communitarian approach developed for the region by Mormons, Powell suggested organizing the land into irrigation and grazing districts that would enable farmers to share water. Despite favorable press coverage, Congress refused Powell's proposals, preferring to cling to the popular myth of the West as a bountiful garden rather than address the area's water shortages.

References: Bartlett 1962; Cohen 2002; Stegner 1953; Wheat 1957–1963; Worster 2001

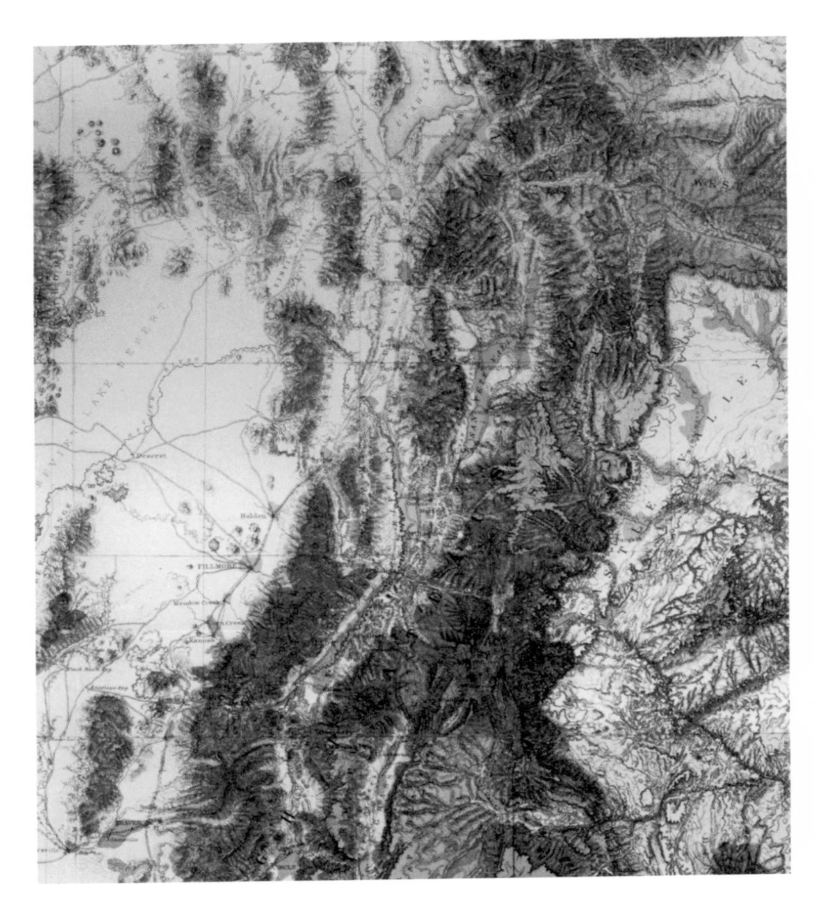

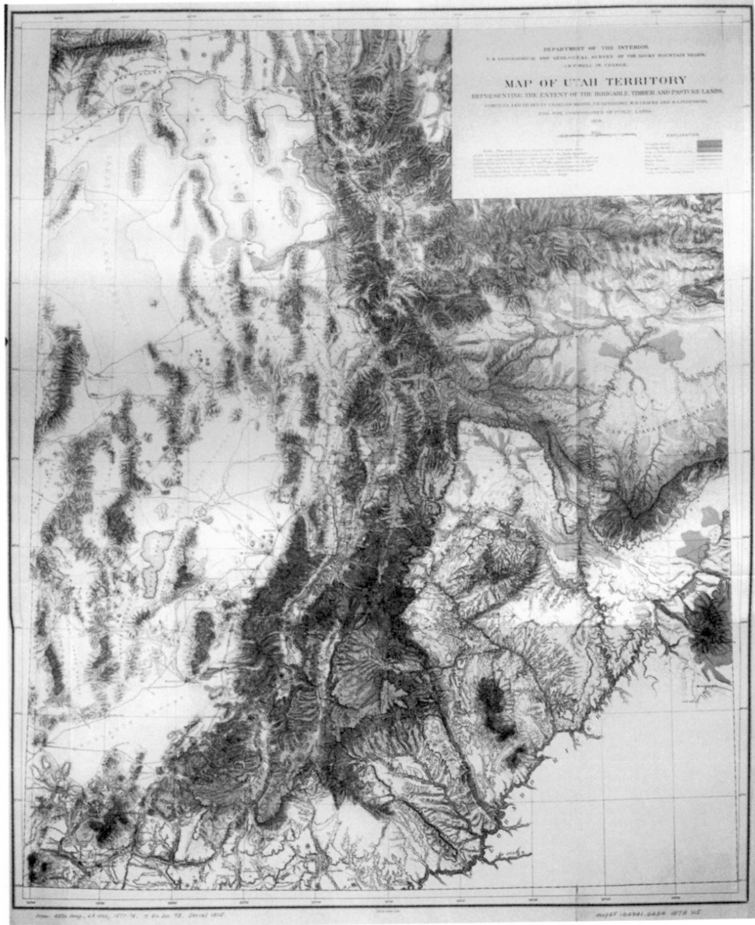

Mapping West of the One Hundredth Meridian

CREATORS: U.S. Army Corps of Engineers, U.S. Geographical Surveys West of the One Hundredth Meridian, George Montague Wheeler (cartographer)

TITLE: *Sketch Indicating the Advancement of the Surveys of the Public Lands and the Military Topographical and Geographical Surveys West of the Mississippi*, 1879, from George M[ontague] Wheeler, *Report of the United States Geographical Surveys West of the One Hundredth Meridian* (Washington, DC: Government Printing Office, 1889)

SIZE: 32.5 x 44.5 inches

LOCATION: Newberry Map 6F G4051 B5 1879 U5

George Montague Wheeler

Dictionary of American Portraits, p. 665.

CREDIT: Newberry Ref oN 7593 C53

George M. Wheeler (1842–1905), a first lieutenant in the Army Corps of Engineers, superintended another of the post-Civil War government surveys, focusing on the area west of the one hundredth meridian. Of the four survey leaders, Wheeler concentrated most intently on mapping. Inspired by the systematic maps of European nations, Wheeler wanted to map the entire United States on a uniform scale in 210 sheets. For the land encompassed by his survey, Wheeler proposed creating ninety-five sheets, mapping the area on a scale of four miles to the inch.

His 1879 sketch summarized the mapping of the American West to date, the numbered rectangles referring to the individual sheets of his partially-completed survey. The land in the southwest that Wheeler himself surveyed or supervised the mapping of is highlighted in tan. To the north, also shaded in tan, is the swath surveyed by Clarence King along the 40th Parallel, following the routes of the Union Pacific and Central Pacific Railroads.

The map's most conspicuous feature is the red grid of six-mile-square townships mapped by the General Land Office (G.L.O.) to record property sales. The federal government hired local surveyors to chart each township's basic topographical features on manuscript maps, sending copies to Washington as well as to local land offices. Because the G.L.O. first mapped areas in demand, the grid indicates the geographic range of the Western real estate market in 1879. By 1890, the G.L.O. had surveyed most of the area represented on the map, but the more detailed topographic mapping Wheeler envisioned would not be realized until the Army Map Service completed the project in 1965.

References: Bartlett 1962; Karrow 1986; Karrow 2002b

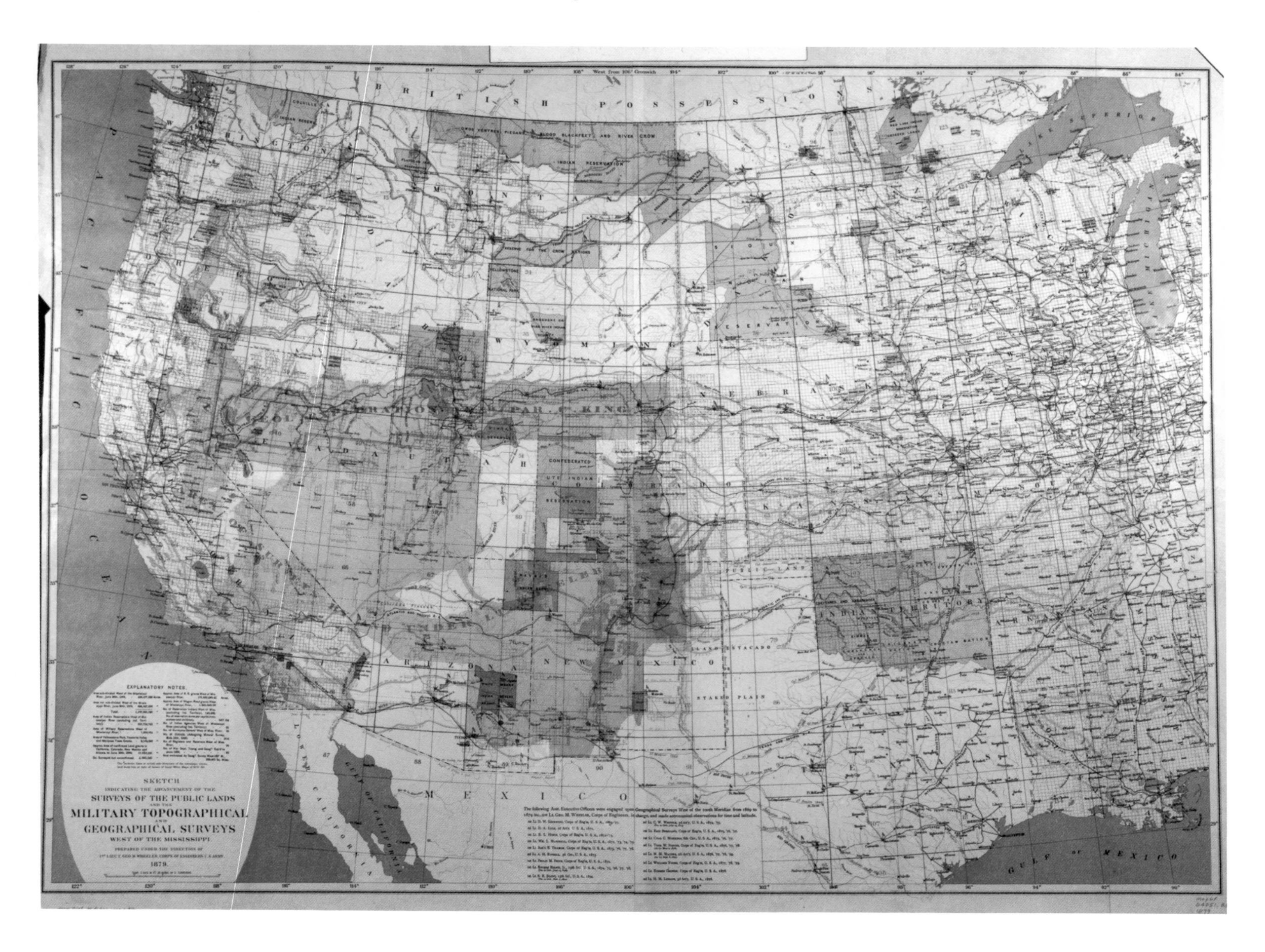

2.11, 2.12

Charting the Pacific Coast

[2.11] CREATORS: United States Coast Survey, James Alden (reconnaissance commander), George Davidson (astronomical observations), Isaac Ingalls Stevens (verification), H. C. Evens (engraving), W. B. McMurtrie (sketch work of shore), Max Strobel (engraving), H. M. Knight (lettering), Frs Herbst (reduction for engraving), E. Hergesheimer (reduction for engraving)

TITLE: *Harbor of San Luis Obispo,* part of *Reconnaissance of the Harbors of Santa Cruz, San Simeon, Coxo and San Luis Obispo, California* ([Washington, DC]: The Survey, 1852)

SIZE: 7.7 x 5.7 inches (part shown, from larger sheet)

LOCATION: National Oceanic and Atmospheric Administration Central Library, Washington, DC
[not on display]

[2.12] CREATORS: United States Coast and Geodetic Survey [F. M. Thorn, L. A. Sengteller (engraver), E. H. C. Lautze, Edward David Taussig]

TITLE: *San Luis Obispo Bay and Port Harford* (Washington, DC: The Survey, 1889)

SIZE: 16 x 26 inches

LOCATION: Regenstein Library, University of Chicago, Chicago, IL
[not on display]

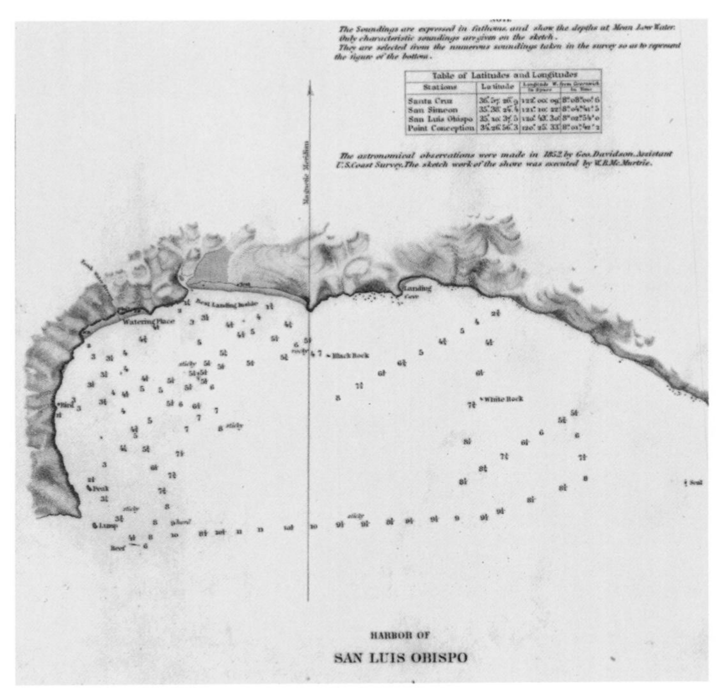

[2.11]

The systematic mapping of the nation's coasts began in 1807, when Thomas Jefferson appointed Ferdinand Hassler, a scientifically-trained Swiss immigrant, to chart the eastern seaboard. Like the Lewis and Clark expedition, Hassler's undertaking aimed to further commercial interests by improving navigation, but rigorous trigonometric methods lent the survey an aura of pure science. The survey expanded as the nation acquired new coastal territories along the Gulf of Mexico and the Pacific Ocean. In 1871, the agency turned inward to tie the surveys of the east and west coasts together with a band of geodetic triangles along the thirty-ninth parallel. This project eventually created the main spatial frame of reference for the entire country.

The coast survey's first map of San Luis Obispo Bay appeared in 1852, two years after the charting of California began. A comparison of the 1852 and 1889 charts illustrates the maturation of coastal mapping. By 1889, the earlier sheet's impressionistic hachures have been replaced by more exact topographic contours. Like Graham's map of Chicago Harbor (see Item 2.15), the San Luis Obispo charts provide depth measurements and identify obstructions with symbols to facilitate the movement of ships. The 1889 map also highlights commercial networks. It locates the Port Harford Wharf on the sheltered leeward side of Point San Luis, which served as a transfer point for the transportation of California's agricultural products. The Pacific Coast Railroad, built in 1874–76 by John Harford, connected the wharf to the city of San Luis Obispo, seven miles inland.

References: Cloud 2007; Edney 1986

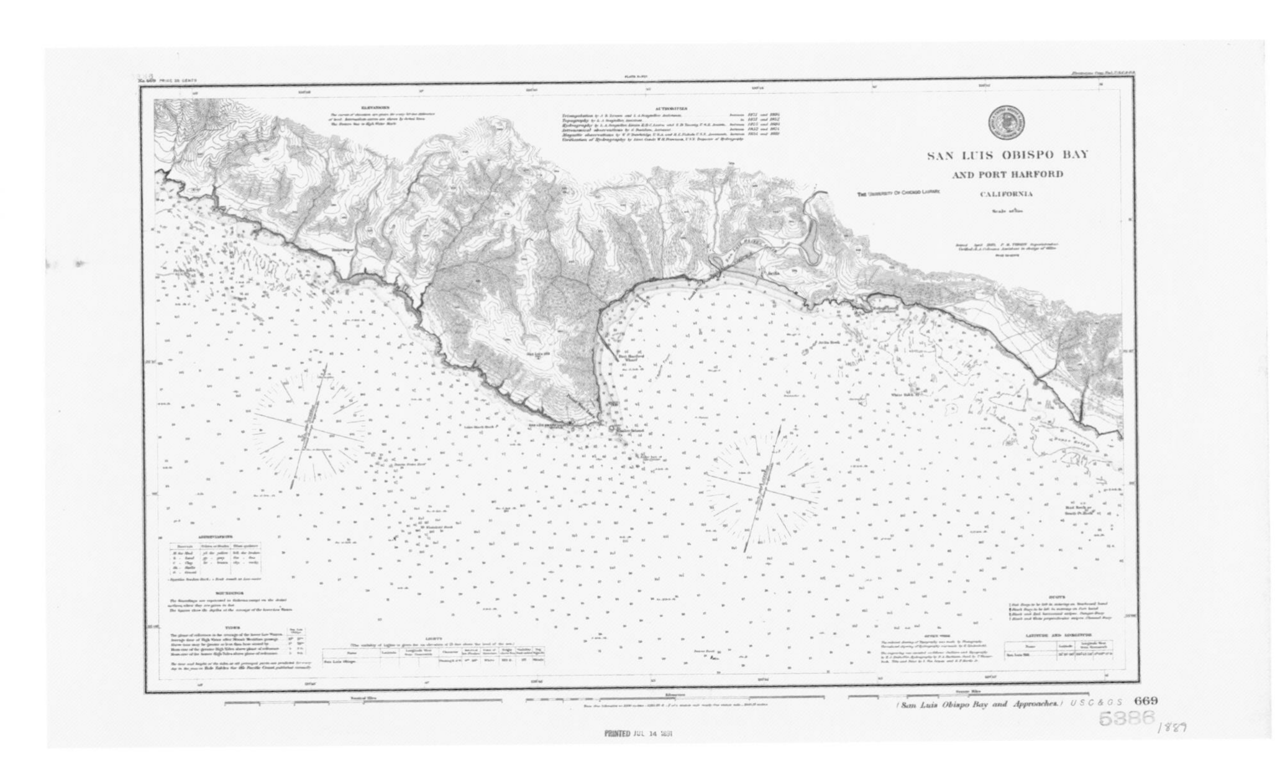

[2.12]

Chicago Becomes Official

Chicago developed on land that had been occupied intermittently by a succession of Native American groups. The Miamis had constructed villages along the Chicago and Des Plaines Rivers in the mid-seventeenth century, and within a few decades the Illinois established hunting camps in the area. By the mid-eighteenth century, the Potawatomis, Ottawas, and Chippewas had settled along the Chicago, Des Plaines, and Kankakee Rivers. The French maintained a colonial presence in the Illinois Country from the 1673 expedition of Jolliet and Marquette until the end of the Seven Years War in 1763, when France ceded the territory to Great Britain.

The land became part of the United States with the 1783 treaty that finalized the nation's independence, but the British lingered in the area for more than a decade, and Native Americans continued to control much of the territory. The balance of power shifted with the defeat of the Indians at the Battle of Fallen Timbers in 1794. The following year, as part of the Treaty of Greenville, the Indians ceded much of Ohio, as well as "one piece of land six miles square at the mouth of the Chikago River, emptying into the south-west end of Lake Michigan, where a fort formerly stood." In 1803, the United States Army erected Fort Dearborn on the strategic site; it soon became a nucleus for the community.

A series of federally-sponsored projects shaped Chicago's phenomenal growth during the nineteenth century. Plans to construct a government-financed canal connecting the Great Lakes to the Mississippi River sparked the city's first boom. To facilitate commercial water traffic, the U.S. Army Corps of Topographical Engineers improved Chicago's harbor and surveyed the Great Lakes between 1841 and 1883. Government engineers drew maps to plan and to record these public works, while the urban growth they inspired led private cartographers to produce updated maps of the city.

2.13, 2.14

Chicago's First Real Estate

[2.13] CREATOR: James Thompson

TITLE: *Plat of Chicago*, 1830 [late nineteenth-century lithograph copy of lost original]

SIZE: 14.8 x 9.6 inches

LOCATION: Chicago History Museum, Chicago, IL [not on display]

[2.14] CREATOR: Joshua Hathaway, Jr.

TITLE: *Chicago with the School Section, Wabansia, and Kinzie's Addition*, (New York: Peter A. Meisner, 1834)

SIZE: 28.8 x 20 inches

LOCATION: Newberry vault Graff 1817

CREDIT: Gift of Edward D. Graff

James Thompson drew Chicago's first legal boundaries in 1830 in preparation for the construction of the Illinois and Michigan Canal. After the federal government subsidized the project with a public land grant, the Illinois General Assembly appointed commissioners to plat towns and sell lots along the route, using proceeds to finance the canal's construction. Surveyor and civil engineer Thompson hailed from the same downstate county as one of the commissioners, which probably helped him get the job in Chicago. Thompson laid out 58 blocks surrounding the fork in the Chicago River, following the spatial grid established by the Land Ordinance of 1785. By most accounts, he filed the plat on August 4, 1830. Auctions of the land began the following month, with 126 lots selling for an average of $35. To the west of the platted area, two parcels of eighty acres were auctioned at $1.25 per acre, and a third parcel sold for a few cents per acre more.

As the eastern terminus of the canal, Chicago began its rise from a swampy frontier outpost to the largest metropolis of the continental interior. The anticipation of canal business sparked a property boom in the 1830s. Feverish real-estate speculation sent land values skyward, and the population swelled from a few hundred in 1830 to over 4,000 by 1837.

Speculators quickly subdivided additional land, creating demand for updated maps. To facilitate the sale of his parcels, John H. Kinzie commissioned Joshua Hathaway to produce the first revised plat, incorporating the blocks Thompson had surveyed within a broader area. When New York lithographer Peter A. Meisner printed 600 copies of Hathaway's map in 1834, *Chicago with the School Section, Wabansia, and Kinzie's Addition* became the first published map of the city. After the economic panic of 1837 ended the real-estate frenzy, more than a decade would elapse before another map of the entire city appeared.

References: Danzer 1984a; Holland 2005; Mayer and Wade 1969

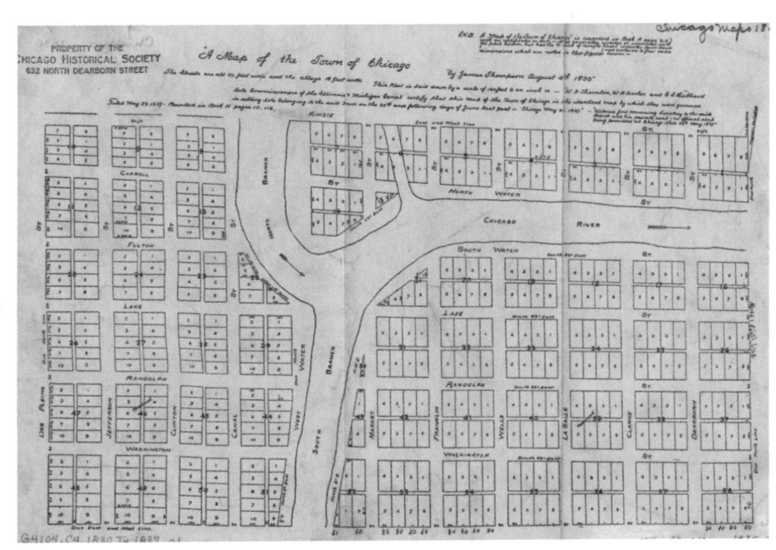

[2.13]

[2.14] DETAIL

[2.14]

Improving Chicago's Harbor

CREATORS: U.S. Army Corps of Engineers, Lieutenant Colonel James Duncan Graham (cartographer), Julius Bien (lithographer)

TITLE: *Chicago Harbor and Bar. Illinois. From Survey made between the 17th of August & the 2nd of September 1858* (New York: Lith. of J. Bien, 1858), from U.S. 36th Congress, 1st Session, 1859–1860. Senate Executive Document 2, 1858

SIZE: 40.4 x 31.2 inches

LOCATION: Newberry Map 6F G4104 C6 2C5 1858 U5

Though Chicago's port on Lake Michigan was pivotal to the city's development as a center of trade and shipping, the mouth of the Chicago River did not provide a reliable natural harbor—a substantial sandbar made the water difficult to navigate and too shallow for large ships to enter. Moreover, the lake's shifting water levels, storms, and wind constantly altered the sandbar's size and shape. The federal government initiated a series of public works to alleviate the problem. Between 1816 and 1828, soldiers at Fort Dearborn dug channels through the bar, which quickly filled again with sand. In 1833, Congress appropriated funds to improve the harbor, resulting in the construction of piers to create a protected channel through the sandbar. The harbor still required constant dredging to remove the ever-accumulating sand, which rose to dangerous levels when federal funds withered.

Lieutenant Colonel James Duncan Graham (1799–1865), of the Army Corps of Engineers, arrived in Chicago in April 1854 to direct harbor improvements and ordered a series of detailed surveys to further the work. His 1858 map of the Chicago Harbor and Bar records the data from soundings taken by the engineers in order to determine the water depth and hence the silt accumulation on the harbor's floor. Graham included detailed notes on the map describing the process he used to calculate the precise latitude and longitude of the site, which enabled him to pinpoint the underwater locations needing to be dredged. Because silt tended to clog the river as well, Graham planned excavations along its south bank. The map enumerates the individual lots to be affected by the project. Graham spent a decade supervising harbor improvements on the Great Lakes, where he was the first to record the existence of lunar tides in 1858–59.

Reference: Holland 2005

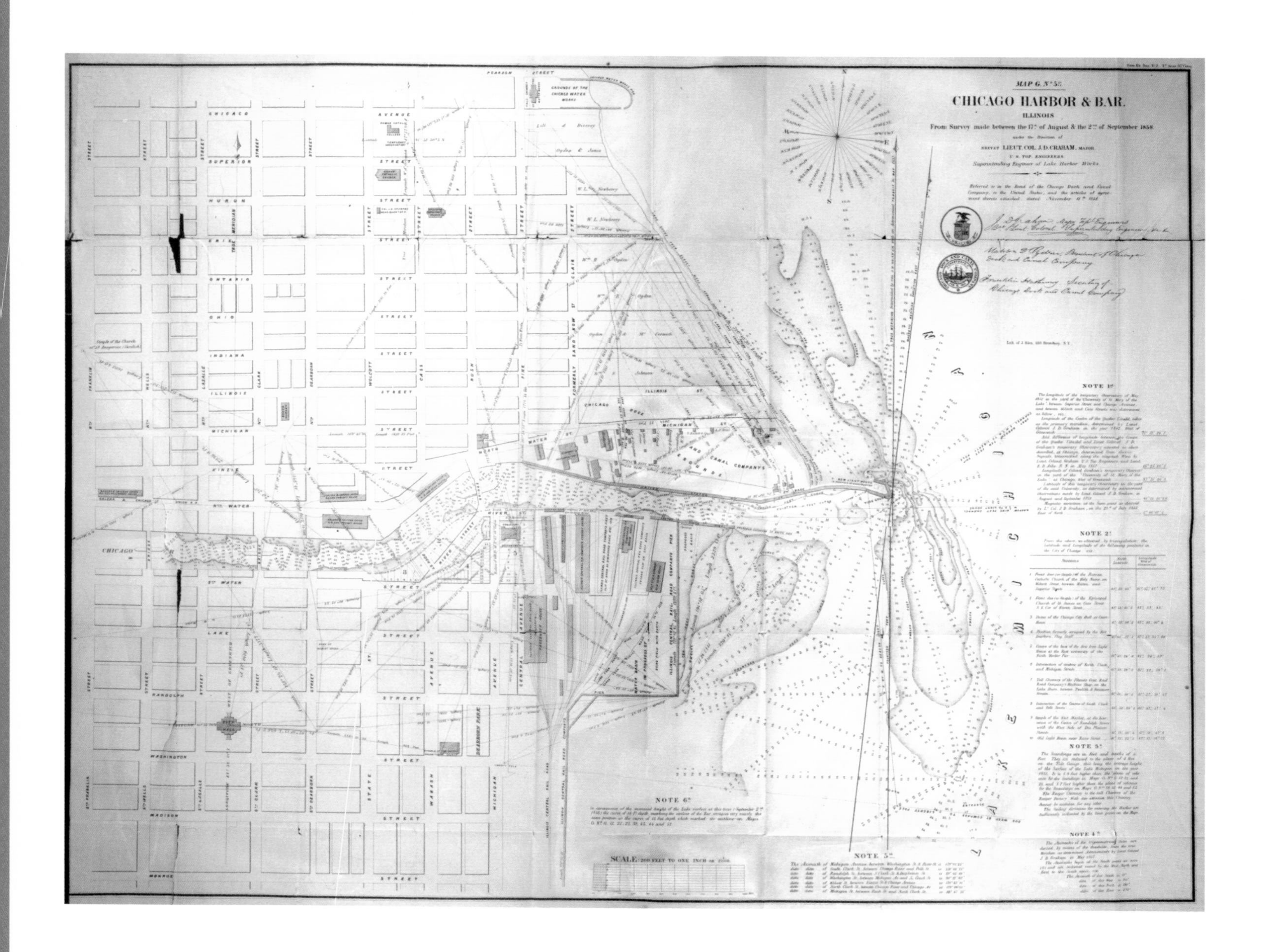

Charting Chicago's Growth

CREATORS: United States Geological Survey, Henry Gannett (chief topographer), John H. Renshawe (geographer), U.S. Lake Survey (triangulation and shoreline), D. C. Harrison, R. C. McKinney, Nat Tyler, Jr., and the Chicago Sanitary Commission (topography)

TITLE: *Chicago Quadrangle* (Washington, DC: United States Geological Survey, 1902)

SIZE: 21 x 18 inches

LOCATION: Newberry Case +G 10896.9 (sheet 3 of 12)

In 1879, the government created the United States Geological Survey (USGS) to consolidate the four Western survey projects. Initially charged with mapping the public domain, the agency's second director, John Wesley Powell, broadened the mandate in 1882 to include creation of a geological map of the entire country. He envisioned the preparation of some 2,600 sheets, in effect forming a detailed national atlas. Powell planned to map cities on the largest scale, one mile to the inch; for areas like the prairies, where the topography was simple, the scale would be four miles to the inch; for mining districts, two miles to the inch. The USGS established a system of dividing the areas to be mapped into quadrangles bounded by universal latitude and longitude lines. The agency's best-known maps are the topographical sheets in the quadrangle series, which continue to be produced today. These "quad sheets" have been put to a variety of uses, from routing highways and utility lines to locating dams and industrial facilities.

The Chicago quad sheet published in 1902 documents Chicago's meteoric growth over the course of the nineteenth century, from a few buildings along the river to a city encompassing 190 square miles. A team of topographical engineers (D. C. Harrison, R. C. McKinney, and Nat Tyler, Jr.) worked with the Chicago Sanitary Commission to survey Chicago in 1889, 1897, and 1899. They mapped natural features such as streams and lakes, man-made additions including railroads and cribs, and political boundaries such as township lines, while the U. S. Lake Survey charted the shoreline. The quad sheet visualized Chicago's increasing density as well as the expansion of its boundaries. Small rectangles denoted individual buildings; bold outlining identified blocks that were completely built up.

References: Edney 1986; Thompson 1981

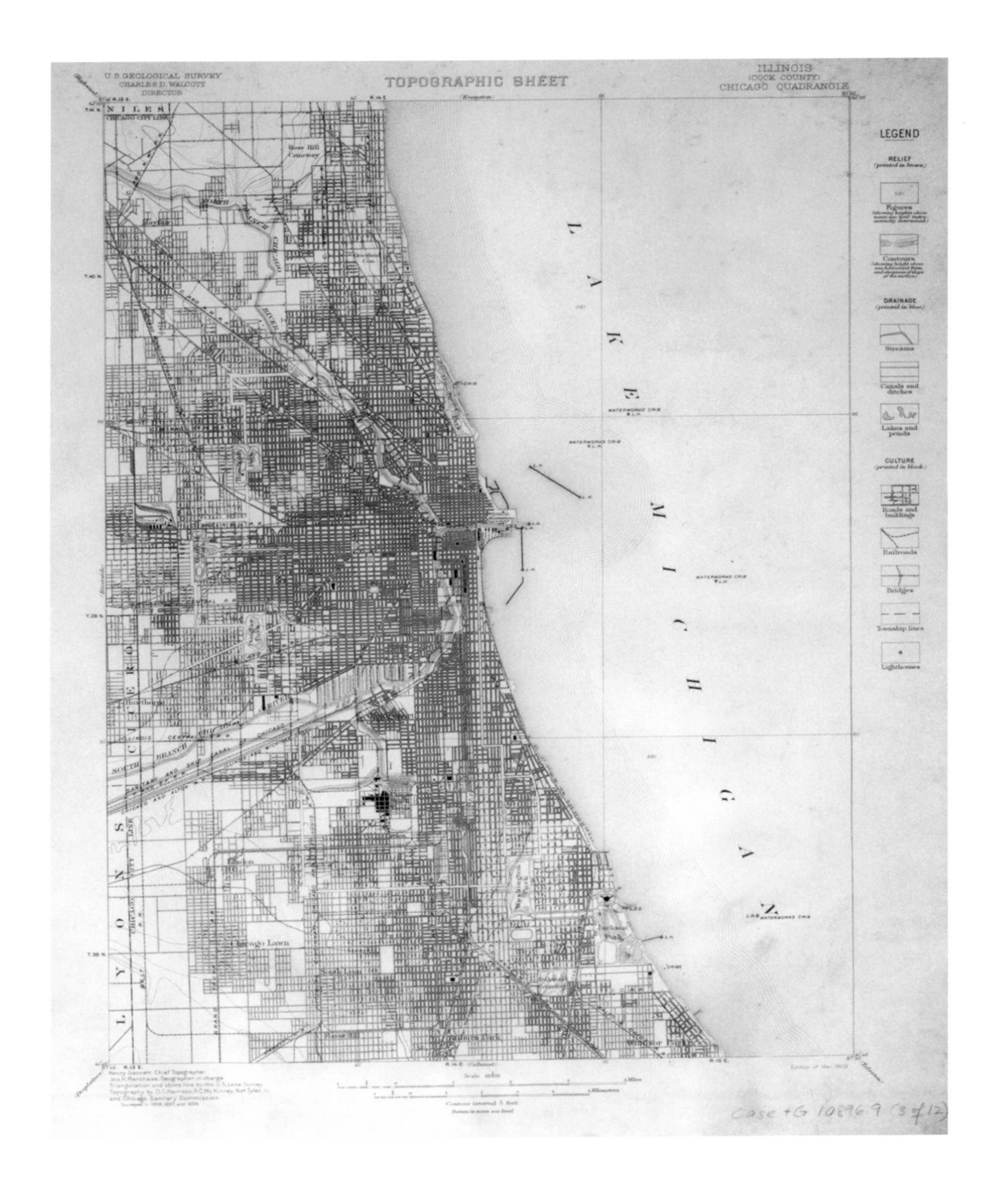

SECTION THREE: MAPPING FOR ENLIGHTENMENT

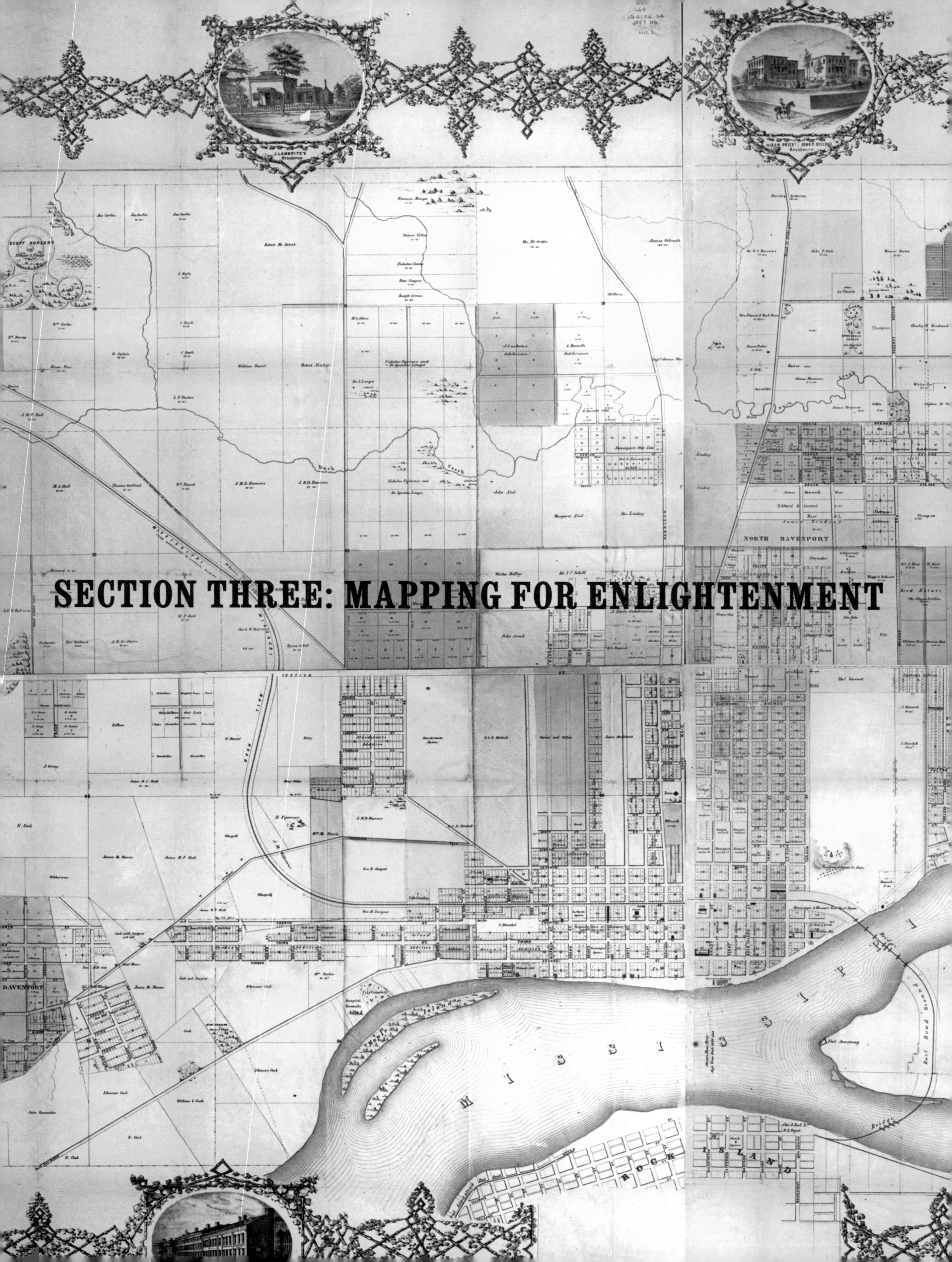

As knowledge of the West's geography deepened over the course of the nineteenth century, Americans improved their understanding of the region through maps and atlases designed for general readers. An expanding segment of the commercial map trade specialized in cartographic products for school and home use. These publications integrated current geographic knowledge about the West into broader surveys of national or world geography. Rarely original in their information, subject matter, or form, the maps broke new ground in their ability to communicate the latest findings and innovations of other cartographers to a wide audience.

Moreover, producers of general-interest maps and atlases often found inventive ways to market their goods. To bolster sales, map publishers had to persuade consumers that geography was important and relevant to their lives. As the settlement and economic development of the West progressed, so too did the public's curiosity about the region. Mapmakers capitalized on this trend by featuring information about the West in popular publications and developing specialized cartographic wares for Western audiences. These maps and atlases in turn stimulated further interest in the region. Because general-interest maps of the West often served the interests of boosters, the line between educational and persuasive cartography was a fine one.

Teaching the West

Generations of Americans saw their first maps of the West in school. Classroom wall maps showed the region within the entire continent, while geography textbooks and school atlases included more specialized maps of individual states and territories. Teachers who regarded geography as an essential discipline for children to master were eager to use these volumes in their classrooms. Furthermore, maps were particularly well suited to prevailing pedagogical methods centered on memorization.

In the second quarter of the century, textbook authors began using thematic maps to illustrate the geographic dimension of social and political phenomena. These diagrams appeared in educational publications and carried cartography's scientific veneer; they seemed objective. But all maps are biased in some way, and these often reproduced popular stereotypes about the region and its people, especially in their representations of Native Americans.

3.1, 3.2

Maps as American History

[3.1] CREATORS: Emma Willard (cartographer), Samuel Maverick (engraver and printer)

TITLE: *Introductory Map: Locations and Wanderings of the Aboriginal Tribes,* in *A Series of Maps to [Accompany] Willard's History of the United States, or Republic of America: Designed for Schools and Private Libraries* (New York: White, Gallaher & White, [1828])

SIZE: 9.75 x 11.5 inches

LOCATION: Newberry Case folio G1201 S1 W5 1828

CREDIT: Bequest of Hermon Dunlap Smith

[3.2] CREATORS: Emma Willard (cartographer), Samuel Maverick (engraver and printer)

TITLE: *Eighth Map: 1789,* in *A Series of Maps to [Accompany] Willard's History of the United States, or Republic of America: Designed for Schools and Private Libraries* (New York: White, Gallaher & White, [1830])

SIZE: 10.13 x 14.38 inches

LOCATION: Newberry Case folio G1201 S1 W5 1830z

CREDIT: Gift of Jameson & Phillips

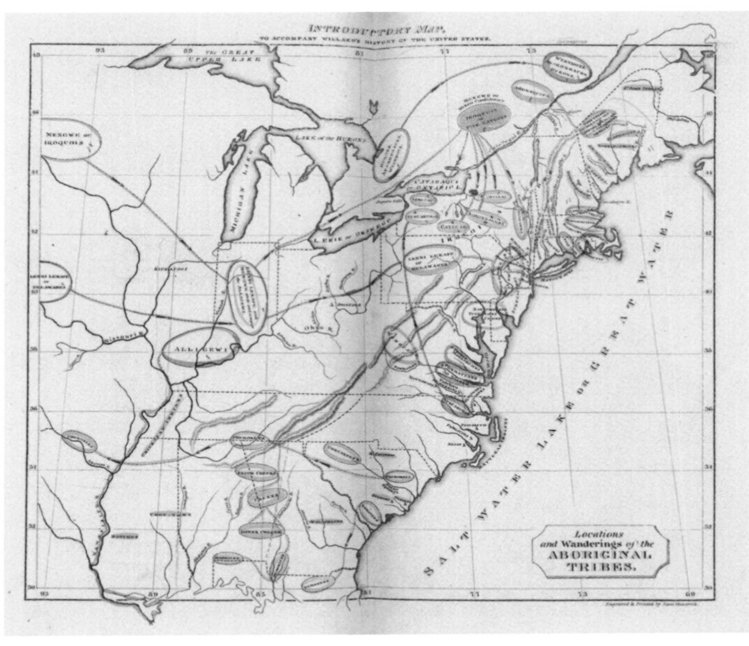

[3.1]

Emma Willard

Henry Barnard, *Educational Biography, Memoirs of Teachers, Educators* (1859): opposite p. 125.

CREDIT: Newberry 141.074

Pioneering educator Emma Hart Willard (1787–1870) founded three schools for girls and authored several history and geography textbooks. The widely-used *Willard's History of the United States* went through nine new editions between 1828 and 1868. To accompany the text, Willard published a series of maps, which she considered "an essential part" of the work, to "be constantly kept before the reader as he proceeds with the narrative." Each map highlighted geographical information pertinent to a specific period in American history.

The atlas dramatizes contemporary views about the position of Native Americans in the evolving nation. On the map *Locations and Wanderings of the Aboriginal Tribes,* depicting the continent before the arrival of Europeans, Indians dominate the continent. Willard used different colors to show the connections between groups, diagramming a complex political and social structure. The arrows tracing migrations, often over long distances, indicates that the natives could claim vast swaths of territory by prior occupation. The *Eighth Map*, representing the United States in 1789, offers a strikingly different view. Willard again employs colorful outlining for political boundaries, but now uses it exclusively to show the borders of the thirteen new states. The Western section of the map appears virtually unoccupied, an empty continent awaiting the expansion of the new nation. Native Americans have been completely erased from the landscape.

In her textbook, Willard subsequently observed that of the many Indians originally inhabiting the eastern coast "most have been exterminated, or driven westward." She went on to survey the numbers and locations of those that remained, both in the East and in the Louisiana territory purchased in 1803.

References: Lutz 1929; Patton 1999; White 1994

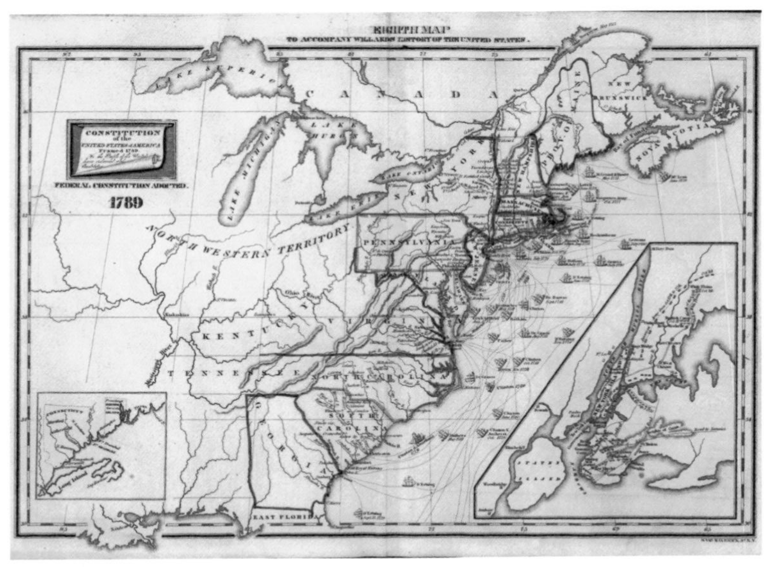

[3.2]

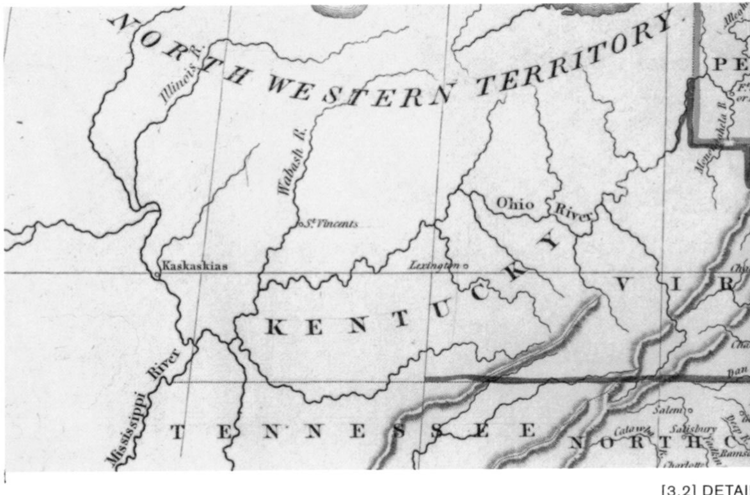

[3.2] DETAIL

The "Barbarous" West

CREATOR: William Channing Woodbridge

TITLE: *Moral and Political Chart of the Inhabited World,* in William C[hanning] Woodbridge, *Modern Atlas: On a New Plan, to Accompany a System of Universal Geography* (Hartford: Beach & Beckwith, 1835)

SIZE: 10.5 x 17 inches

LOCATION: Newberry Case folio G 10 .979

CREDIT: Gift of Louise St. John Westervelt

William Channing Woodbridge
Henry Barnard, *Educational Biography, Memoirs of Teachers, Educators* (1859): opposite p 268.
CREDIT: Newberry I41.074

William Channing Woodbridge's (1794–1845) geography publications reflected his lifelong dedication to improving education. As a teacher, journal editor, and author, Woodbridge pursued the latest pedagogical and scientific innovations. He was one of the first school-atlas authors to incorporate thematic maps. Woodbridge's *Moral and Political Chart of the Inhabited World* first appeared in his 1826 *School Atlas Designed to Accompany Woodbridge's Rudiments of Geography.* He reprinted the chart in subsequent atlases, including his 1835 *Modern Atlas.*

Using a world map as its base, Woodbridge employed varying patterns of gray shading to rank parts of the globe by their level of civilization, from "savage" to "enlightened." By adding symbols to indicate the type of government and religion in each area, Woodbridge suggested that these forms of social organization determined each culture's degree of civilization. He underscored the racial implication of his system by using the darkest hatching to color the "primitive" and "barbarous" regions and the palest pattern, in a form evoking radiant light, to mark the "enlightened" areas.

The chart boldly contrasts eastern and western North America. Woodbridge identified the residents of the United States as Republican, Christian, and enlightened, whereas the vast Western territory is filled with "unsubdued Indians"—"pagans" ruled by "independent chiefs or dukes" who remain "barbarous" or "primitive." In Woodbridge's scheme, North America contained more land dominated by "barbarous" people than any other continent.

Reference: Patton 1999

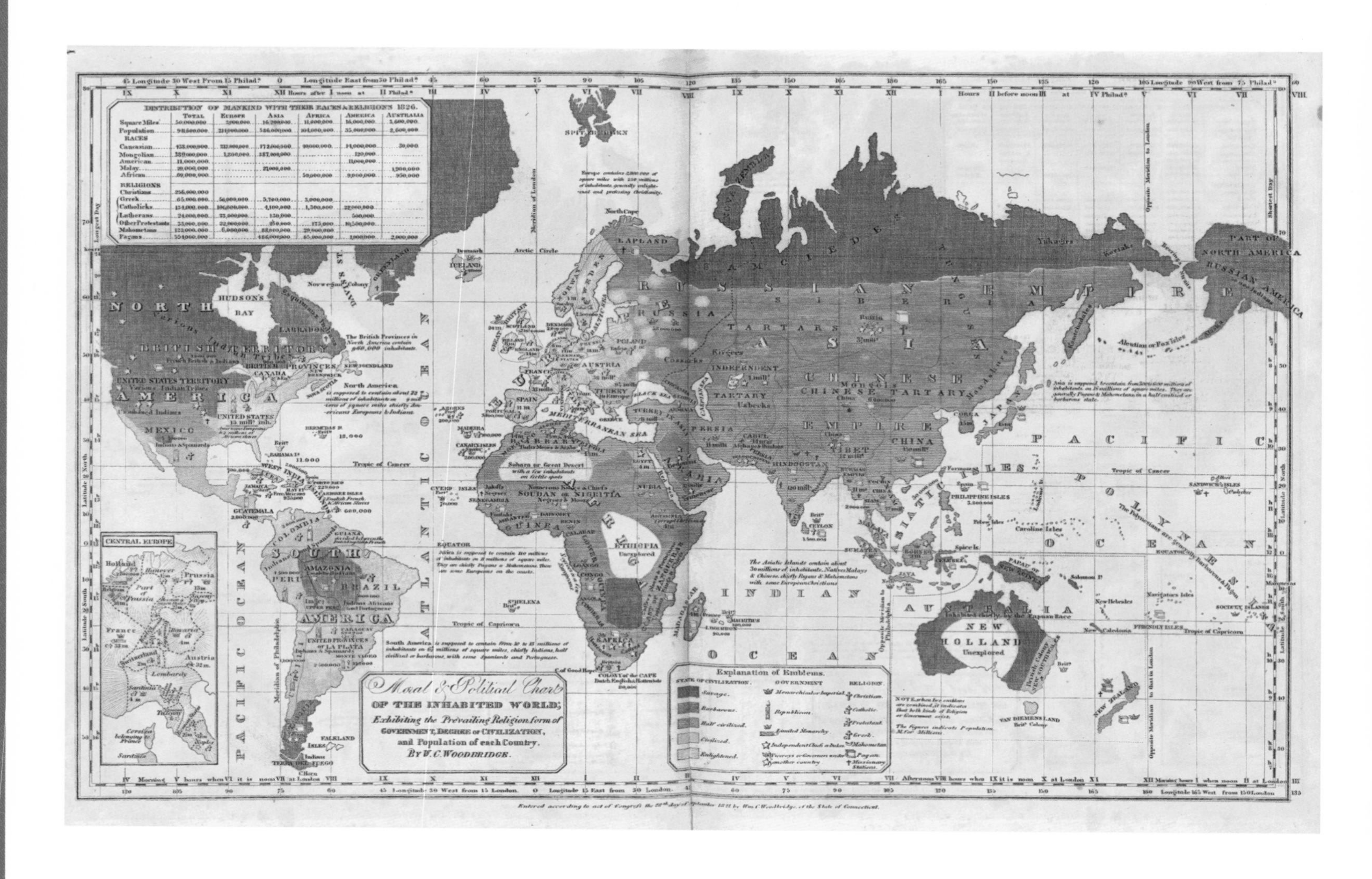

3.4, 3.5

The West as a Land of Promise

[3.4] CREATOR: S. Augustus Mitchell

TITLE: *School Geography: System of Modern Geography…* (Philadelphia: Cowperthwait, Desilver, & Butler, 1854)

SIZE: 6.5 x 7.86 inches (two-page spread)

LOCATION: Newberry Cassidy M5 S34 1854 c

CREDIT: Gift of Martin M. Cassidy

[3.5] CREATOR: S. Augustus Mitchell

TITLE: *Map of the Territories and Texas & California* and *Questions on the Map of the Western Territories* in *Primary Geography* (Philadelphia: Cowperthwait, 1858), pp. 65-66

SIZE: 6.25 x 5.16 inches (map)

LOCATION: Newberry Cassidy M5 P7 1858

CREDIT: Gift of Martin M. Cassidy

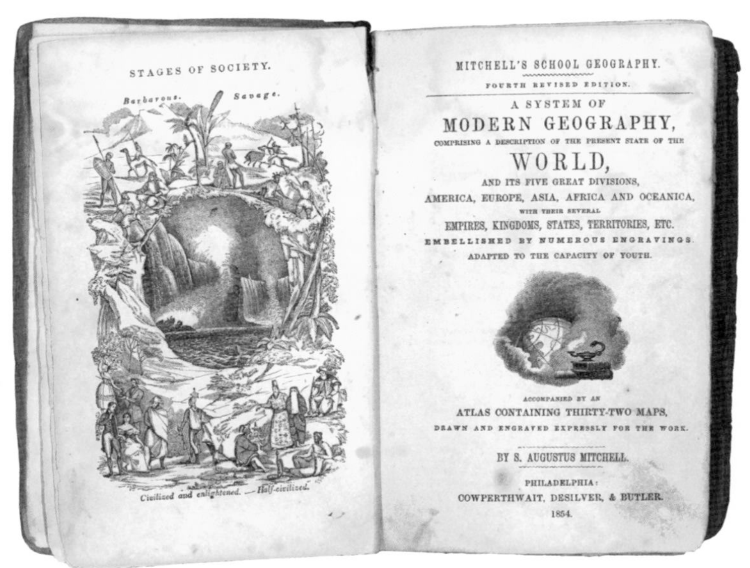

MITCHELL'S SCHOOL GEOGRAPHY.
FOURTH REVISED EDITION.
A SYSTEM OF
MODERN GEOGRAPHY,
COMPRISING A DESCRIPTION OF THE PRESENT STATE OF THE
WORLD,
AND ITS FIVE GREAT DIVISIONS,
AMERICA, EUROPE, ASIA, AFRICA AND OCEANICA,
WITH THEIR SEVERAL
EMPIRES, KINGDOMS, STATES, TERRITORIES, ETC.
EMBELLISHED BY NUMEROUS ENGRAVINGS.
ADAPTED TO THE CAPACITY OF YOUTH.

ACCOMPANIED BY AN
ATLAS CONTAINING THIRTY-TWO MAPS,
DRAWN AND ENGRAVED EXPRESSLY FOR THE WORK.

BY S. AUGUSTUS MITCHELL.

PHILADELPHIA:
COWPERTHWAIT, DESILVER, & BUTLER.
1854.

[3.4]

Connecticut-born Samuel Augustus Mitchell (1792–1868) began his career as a schoolteacher. Frustrated by the low quality of available geographical works, he determined to turn out improved maps and atlases. To pursue his new vocation, Mitchell moved around 1830 to Philadelphia, then the nation's center of commercial publishing and map production. Lacking training in either cartography or engraving, Mitchell served primarily as an editor and business manager for his publishing ventures. For several decades he worked with J. H. Young, the firm's primary map engraver. After building a reputation for wall maps, travel guides, and general atlases, Mitchell published his first school atlas in 1839. The firm, continued by his son, issued educational atlases until 1886.

Mitchell's texts stressed descriptive geography, enumerating the basic topographical features, natural resources, and products for each geographical unit, followed by quiz questions. He presented the information impartially, but the texts were patently hierarchical. Mitchell discussed America first and in the greatest detail, relegating the rest of the world to brief summaries in the books' final pages.

In surveying the United States from east to west, Mitchell gave pride of place to the eastern states, but he cast the West as the land of promise for the future, observing in the 1854 *System of Modern Geography* that "no region of equal extent in the world exhibits such a combination of mineral wealth and fertility of soil, united with such rare facilities of transportation." He went on to note the region's "current of emigration which is filling up the country with unexampled rapidity." Like most of his contemporaries, Mitchell regarded settlement of the West by whites as a key to the region's progress. Native Americans remained a stumbling block, as "nearly all of the Indians in the United States" could be found in the West, where they "retain their original savage character and condition." The book's engraved frontispiece reinforced this impression, contrasting the scantily clad, buffalo-hunting Indians in the "savage" quadrant to the elegantly dressed, decorously posed whites in the "civilized and enlightened" realm.

References: Barnard 1859, Patton 1999

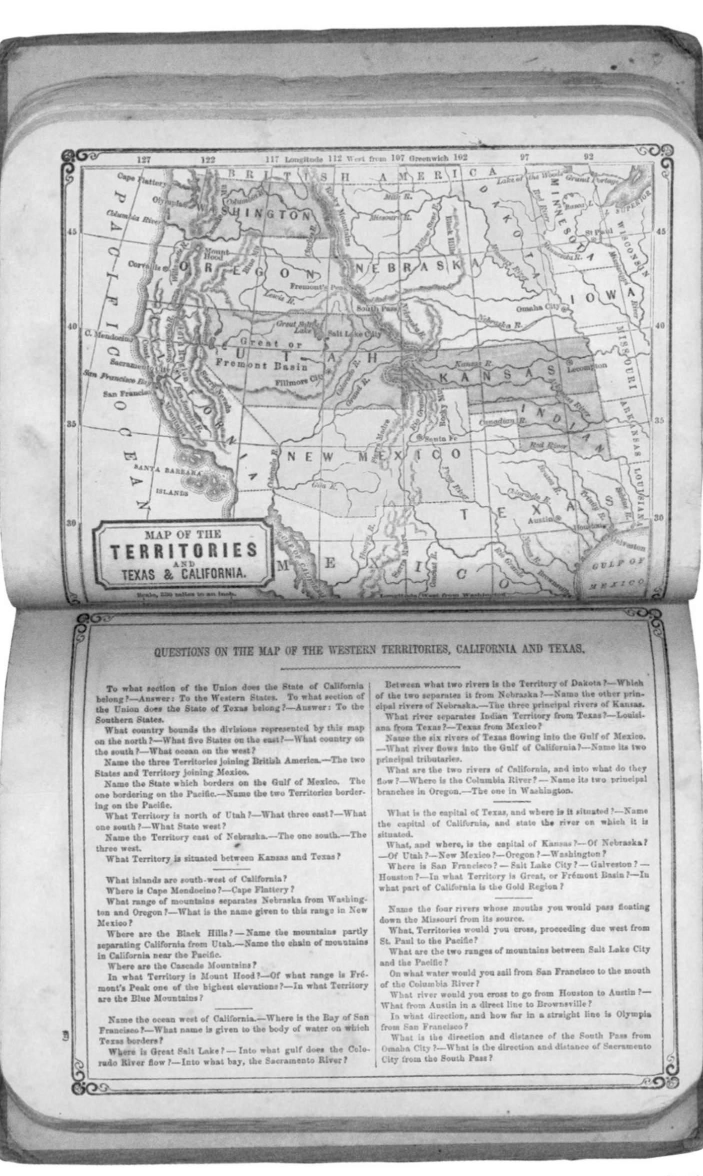

QUESTIONS ON THE MAP OF THE WESTERN TERRITORIES, CALIFORNIA AND TEXAS.

To what section of the Union does the State of California belong?—Answer: To the Western States. To what section of the Union does the State of Texas belong?—Answer: To the Southern States.

What country bounds the divisions represented by this map on the north?—What five States on the east?—What country on the south?—What ocean on the west?

Name the three Territories joining British America.—The two States and Territory joining Mexico.

Name the State which borders on the Gulf of Mexico. The one bordering on the Pacific.—Name the two Territories bordering on the Pacific.

What Territory is north of Utah?—What three east?—What one south?—What State west?

Name the Territory east of Nebraska.—The one south.—The three west.

What Territory is situated between Kansas and Texas?

What islands are south-west of California?

Where is Cape Mendocino?—Cape Flattery?

What range of mountains separates Nebraska from Washington and Oregon?—What is the name given to this range in New Mexico?

Where are the Black Hills?—Name the mountains partly separating California from Utah.—Name the chain of mountains in California near the Pacific.

Where are the Cascade Mountains?

In what Territory is Mount Hood?—Of what range is Frémont's Peak one of the highest elevations?—In what Territory are the Blue Mountains?

Name the ocean west of California.—Where is the Bay of San Francisco?—What name is given to the body of water on which Texas borders?

Where is Great Salt Lake?—Into what gulf does the Colorado River flow?—Into what bay, the Sacramento River?

Between what two rivers is the Territory of Dakota?—Which of the two separates it from Nebraska?—Name the other principal rivers of Nebraska.—The three principal rivers of Kansas.

What river separates Indian Territory from Texas?—Louisiana from Texas?—Texas from Mexico?

Name the six rivers of Texas flowing into the Gulf of Mexico.—What river flows into the Gulf of California?—Name its two principal tributaries.

What are the two rivers of California, and into what do they flow?—Where is the Columbia River?—Name its two principal branches in Oregon.—The one in Washington.

What is the capital of Texas, and where is it situated?—Name the capital of California, and state the river on which it is situated.

What, and where, is the capital of Kansas?—Of Nebraska?—Of Utah?—New Mexico?—Oregon?—Washington?

Where is San Francisco?—Salt Lake City?—Galveston?—Houston?—In what Territory is Great, or Frémont Basin?—In what part of California is the Gold Region?

Name the four rivers whose mouths you would pass floating down the Missouri from its source.

What Territories would you cross, proceeding due west from St. Paul to the Pacific?

What are the two ranges of mountains between Salt Lake City and the Pacific?

On what water would you sail from San Francisco to the mouth of the Columbia River?

What river would you cross to go from Houston to Austin?—What from Austin in a direct line to Brownsville?

In what direction, and how far in a straight line is Olympia from San Francisco?

What is the direction and distance of the South Pass from Omaha City?—What is the direction and distance of Sacramento City from the South Pass?

[3.5]

Studying the West at Home

CREATOR: George Franklin Cram

TITLE: *Dakota*, in George F[ranklin] Cram, *Cram's Unrivaled Family Atlas of the World* (Chicago: George F. Cram, 1883)

SIZE: 14.3 x 12.13 inches (book), 18.75 x 11.75 inches (two-page spread showing *Dakota*)

LOCATION: Newberry Folio G 10 192

CREDIT: Gift of Richard Boothby Stoops

George Franklin Cram
CREDIT: Chicago History Museum

George Franklin Cram (1842–1928) built a successful career as a cartographic entrepreneur in Chicago after the Civil War. He probably moved to Chicago from Lowell, Massachusetts, at the suggestion of his uncle, map publisher Rufus Blanchard, who became his employer and then business partner. In 1869 Cram started his own company. He issued a variety of maps, but is best remembered for his atlases. Cram's *Unrivaled Family Atlas of the World* proved his most enduring product. First published in the early 1880s, new editions of the atlas appeared into the 1950s.

The family atlas complemented the textbooks, encouraging geographical learning in the home. Like the school atlases, Cram's home reference atlas promoted nationalism by discussing the United States first and in the greatest detail. By devoting a full page to almost every state and territory, Cram accorded equal attention to each part of the country and honored state sovereignty. The book's design maximized practical information (particularly place names) and avoided ornamental flourishes that would increase its price. Cram reserved decorative exuberance and Western pride for the title page, whose montage of images can be read as an allegory of westward expansion. The ship at the top right symbolizes European discovery of the continent, while the farm scene depicts the settling of the heartland. The train represents the technological conquest of the frontier, opening the West's sublime mountain scenery and resources to emigrants and developers.

Reference: Danzer 1984b

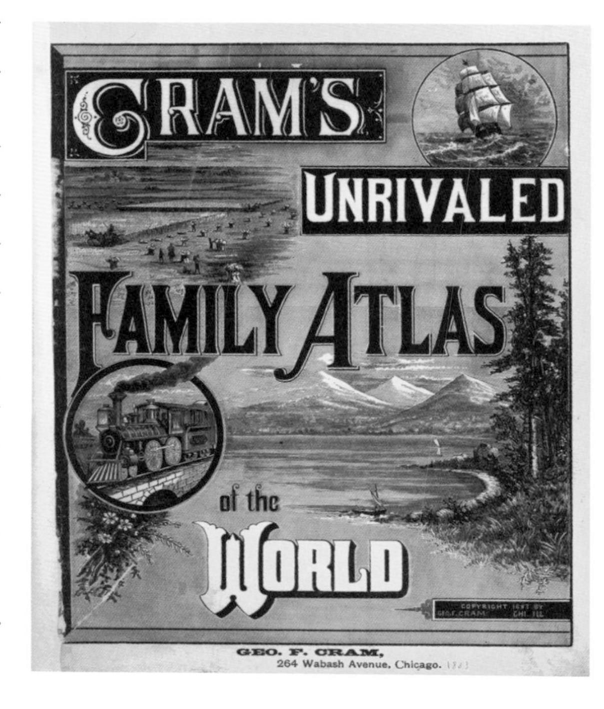

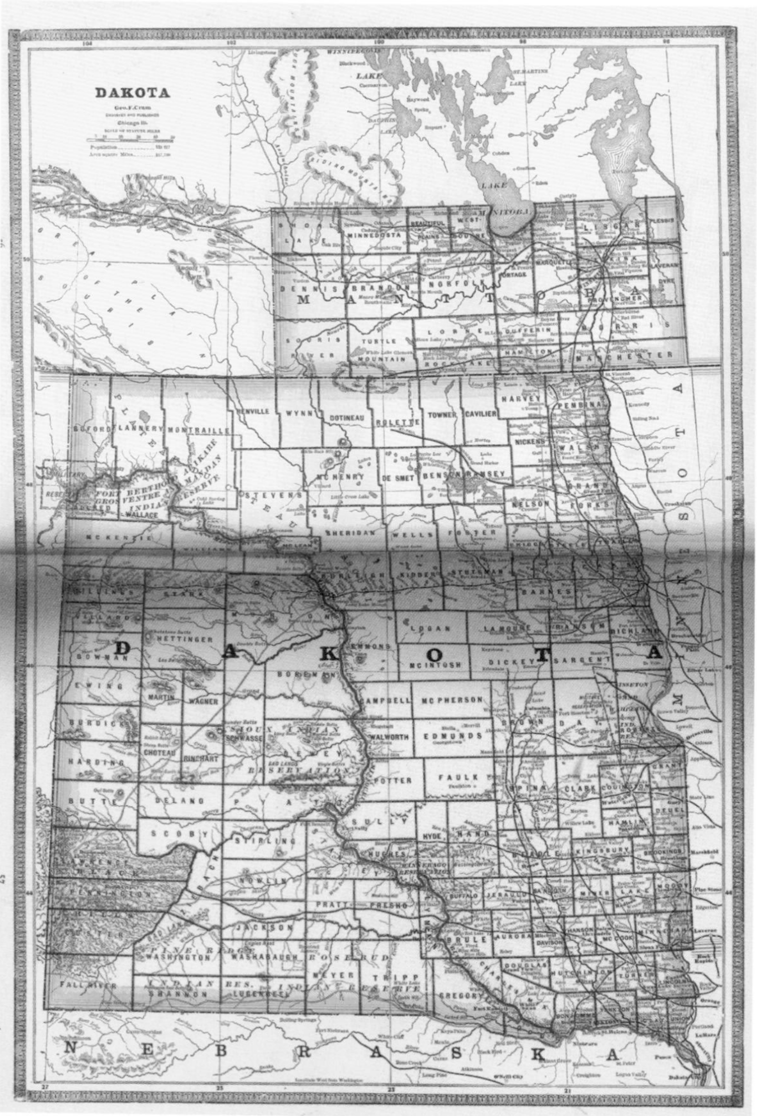

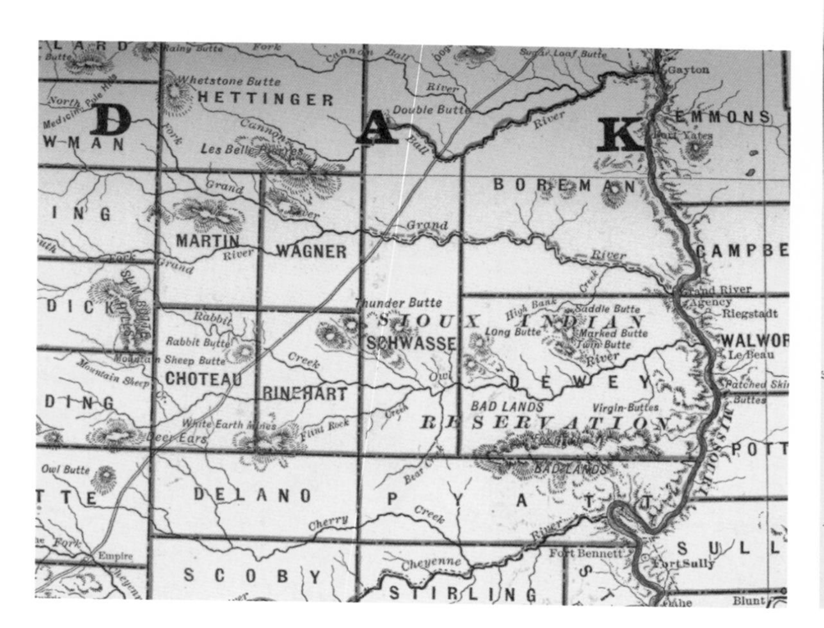

Promoting the West

Privately-made maps picked up where government maps left off in promoting the ideology of expansion and facilitating western settlement. Commercial firms filled gaps, mapping areas not surveyed by government cartographers and producing maps for specialized purposes. They also restyled data from government maps to create attractive regional maps for popular consumption.

Much as Western boosters used maps to sell the region, enterprising cartographers exploited rising popular interest in the West to sell maps. Guidebooks for emigrants, promotional tracts, political pamphlets, and broadsides incorporated maps to further their causes. These cartographic goods advanced the economic development of the West by conceptualizing the region itself as a commodity—as something that could be bought and sold, endlessly repackaged and exchanged.

Luring Settlers to the Midwest

CREATOR: John Mason Peck

TITLE: *Western States,* in J[ohn] M[ason] Peck, *A Guide for Emigrants Containing Sketches of Illinois, Missouri, and the Adjacent Parts* (Boston: Lincoln and Edmands, 1831)

SIZE: 5.86 x 7.25 inches

LOCATION: Newberry Graff 3234

CREDIT: Gift of Everett D. Graff

Frontier intellectual and clergyman John Mason Peck (1789–1858) worked tirelessly to promote the settlement of the Midwest. Raised on a Connecticut farm, ordained as a Baptist minister in 1813, Peck moved to St. Louis the same year and established the First Baptist Church and the earliest missionary society in the West. He continued his religious work after relocating to southern Illinois, where he helped establish Rock Spring Seminary, which became the state's first college.

Peck's emigrant guides, his *Gazetteer of Illinois,* and his important biography of Daniel Boone helped lay the cornerstone of the area's early printed literature. With an eye toward writing a comprehensive history of the region, Peck assembled a vast library of books, periodical articles, newspapers, pamphlets, and reports of religious and philanthropic societies. Fire destroyed the collection in 1852, and though Peck worked to recreate his collection, he never completed his history. For all of his endeavors, Peck's incentives were religious and educational rather than economic. He had no financial stake in the development of the region and probably realized little profit from his publications.

A Guide for Emigrants provided an overview of the Mississippi Valley, paying special attention to Illinois and Missouri. Peck surveyed the topography, summarized agricultural and manufacturing production, and enumerated social and cultural institutions. The map folded into the book's front illustrates the trajectory of the region's development, beginning with the Ohio and Mississippi Rivers. Southern Illinois boasts many more place names than the northern part of the state; Vandalia and Edwardsville are marked in bolder type than Chicago, which had yet to be incorporated as a town.

References: Buck 1914; Hoover 1989

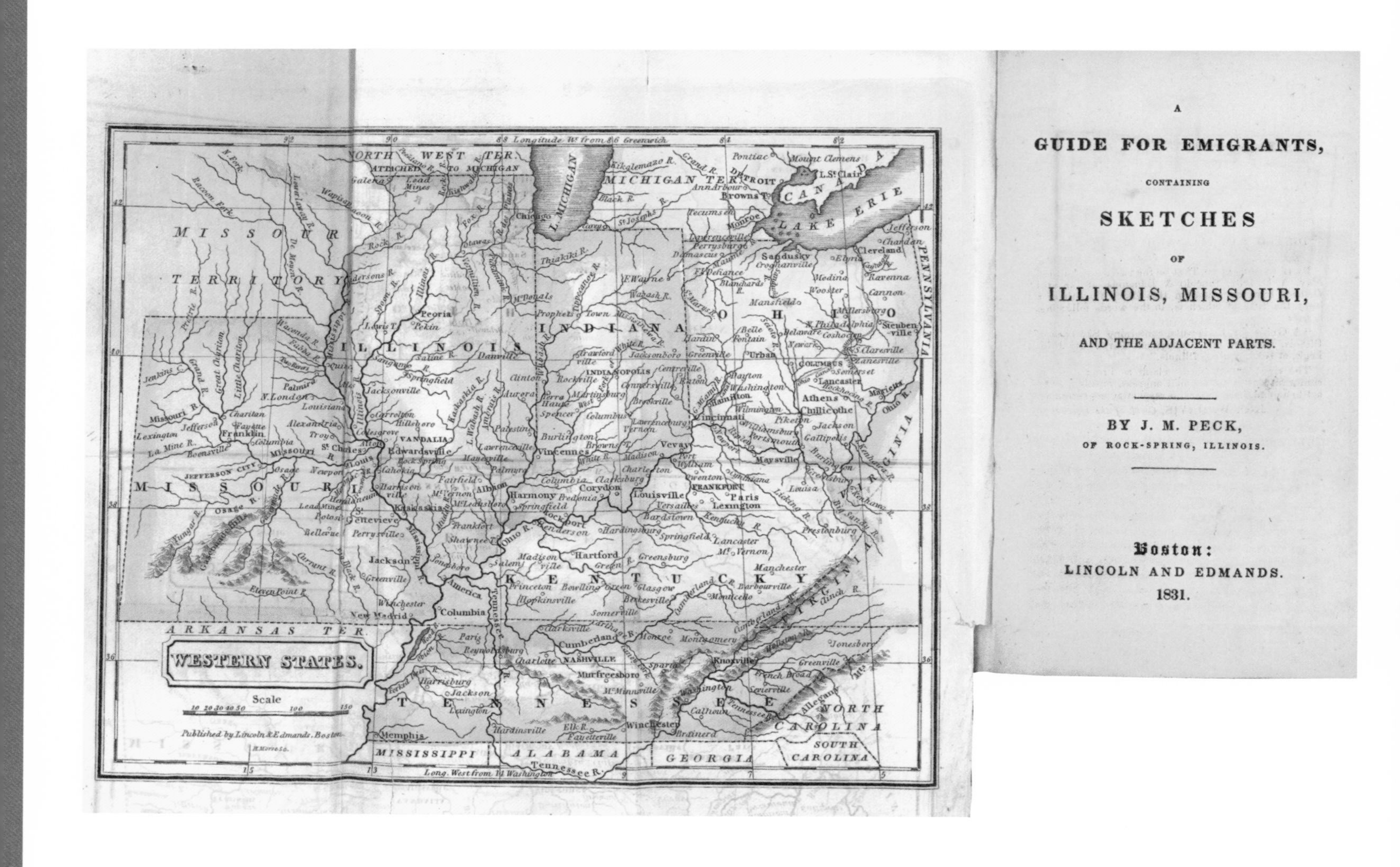

A

GUIDE FOR EMIGRANTS,

CONTAINING

SKETCHES

OF

ILLINOIS, MISSOURI,

AND THE ADJACENT PARTS.

BY J. M. PECK,

OF ROCK-SPRING, ILLINOIS.

Boston:

LINCOLN AND EDMANDS.

1831.

3.8

Negotiating the United States–Mexican Border

CREATOR: John Disturnell

TITLE: *Mapa de los Estados Unidos de Méjico: segun lo organizado y definido por las varias actas del congreso de dicha républica y construido por las mejores autoridades* (*Map of the United States of Mexico: organized and defined by the various acts of the Congress of this republic and constructed according to the best authorities*) (Nueva York: J. Disturnell, 1847)

SIZE: 28 x 57.2 inches

LOCATION: Newberry Graff 1092

CREDIT: Gift of Everett D. Graff

John Disturnell's (1801–1877) motives for printing his *Mapa de los Estados Unidos de Méjico* were unabashedly economic. After establishing a reputation as a publisher of guidebooks to the Northeast and Great Lakes regions, New York-based Disturnell followed the market westward. With this map, he capitalized on the interest sparked by the outbreak of the Mexican-American War in 1846. Disturnell issued at least 23 editions of the map, seven in 1846 alone.

Disturnell did not create his *Mapa de los Estados Unidos de Méjico* from scratch; he merely reprinted, with minor changes, a map issued in 1828 by White, Gallagher and White, who had in turn pirated it from an 1826 edition of Henry S. Tanner's map of North America. Although Tanner had synthesized reputable sources, including maps of New Spain by Alexander von Humboldt and Zebulon Pike, the geographical data was considerably outdated by the time Disturnell recycled it.

When the war ended, the diplomats who negotiated the Treaty of Guadelupe Hidalgo relied on Disturnell's map to establish the new border between the United States and Mexico. Nicholas P. Trist, who composed the treaty, knew little about the geography through which the border would run. He recognized Disturnell's inaccuracies, but time pressures prevented finding a better source. The map's shortcomings contributed to difficulties in determining the southern boundaries of New Mexico and California. The diplomats used at least three different editions of the map, attaching the seventh edition to the American copy of the treaty and the twelfth edition to the Mexican copy. The United States Senate ratified the treaty in March 1848.

The following year, U.S. military explorer Randall B. Marcy enumerated the map's failings: "He makes a greater error than most others in laying down the Pecos, and has the Colorado, Brazos, and Red river all inaccurately placed. Upon the Red river he has a very large branch coming from the far west, near El Paso, which he calls 'Ensenado Choctaw.' This is an altogether imaginary stream..."

Despite its inaccuracies, Disturnell's map looked authoritative, boasting Spanish text, statistical tables, a detailed inset, and a patriotic American eagle. The map became the standard geographical reference for the contested area, prompting at least 20 more versions after 1848. The federal government subsequently aimed to rectify the border by ordering a scientific boundary survey, executed between 1850 and 1857.

References: Goetzmann 1966; Martin 1972; Rebert 2001

3.9

The Destiny of the Heartland

CREATORS: William Gilpin (cartographer), Julius Hutawa (printer), T. S. Wagner (lithographer)

TITLE: *Gilpin's Hydrographic Map of North America.* (St. Louis: Juls. Hutawa Lithr., 1848)

SIZE: 32 x 41.6 inches

LOCATION: Newberry Map 6F G3301 C2 1848 G5

William Gilpin
New England Magazine ns 7 (1892): 256.
CREDIT: Newberry A5.115

Often called the prophet of the American West, William Gilpin's (1813–1894) intellectual creativity, ambition, and enthusiasm matched the grand scale of the region he championed. The Philadelphia native got his first taste of the West when he arrived in St. Louis to recruit soldiers for the Army during the 1836 Seminole War. When he returned to St. Louis after the war, Gilpin began his life's work promoting the West as editor of the *Missouri Daily Argus.* Gilpin initially focused his attention on the geographic centrality of the Mississippi Valley, but expanded his compass to the entire trans-Mississippi after a chance meeting with Frémont on the Oregon Trail during a trip to the Pacific coast in 1843–44. Gilpin went on to become the first governor of the Colorado Territory.

Gilpin trumpeted his ideas through popular lectures, magazine features, and books, all illustrated with maps. In the tradition of Missouri Senator Thomas Hart Benton, he argued that a succession of empires, progressing east to west from Asia to America, had shaped world history. He rationalized this theory with German geographer Alexander von Humboldt's notion of an isothermal zodiac, a band thirty-five degrees wide snaking above and below the fortieth parallel. Observing that history's great empires had all arisen within this zone, Gilpin identified it as the best route for the transcontinental railroad, which would enable the American West to realize its destiny as the next center of civilization.

Gilpin developed his geopolitical ideas about centers and peripheries in his *Hydrographic Map of North America.* By imposing a ring of concentric circles on the map, he marked the continent's geographical center, in present-day Kansas. Gilpin thus diagrammed the continent as a concave structure, united by an interior river system. By pushing the Rocky Mountains west of their actual position, he enlarged the Mississippi Valley's fertile central basin, downplaying the obstacles of mountains and deserts. Coloring the map with washes of green and gold, Gilpin pictured North America as a sunny, salubrious expanse filled with natural resources. He included a small version of his 1848 wall map as the frontispiece to his 1860 treatise *The Central Gold Region.* Gilpin's optimistic vision of national unity and geographic destiny gained new urgency as the Civil War approached.

References: Gilpin 1860; Karrow 2002a; Smith 1973 [1950]

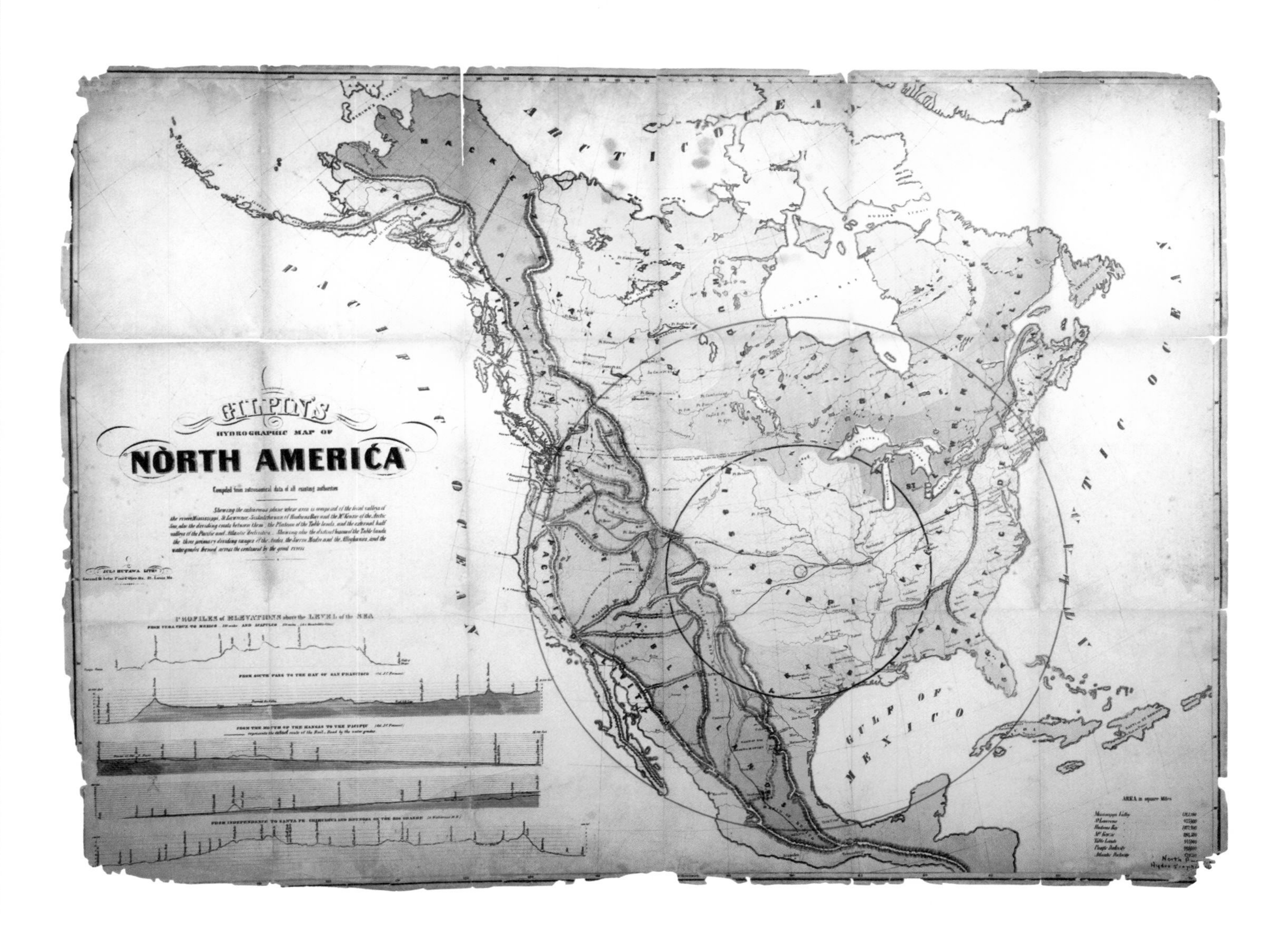

3.10

Squandering the Public Domain

CREATOR: Anon.

TITLE: *Map Showing How the People's Land has been Squandered upon Corporations,* from *The History of Our Public Lands* ([New York]: Justice Publishing Co., c. 1882)

SIZE: 24 x 28.5 inches

LOCATION: Newberry Map 4F 3701 G465

In contrast to the many maps championing westward expansion, the *Map Showing How the People's Land has been Squandered upon Corporations* took a more critical approach. Visualizing the argument of an accompanying pamphlet, *The History of Our Public Lands,* the map highlighted the vast swaths of federal territory granted to railroads to subsidize the building of a national transportation network, as well as the millions of dollars given them in loans. Railroads typically received every other 640-acre section in a band twenty miles wide along their proposed route, creating a checkerboard pattern of railroad- and government-owned parcels. In total, the firms received more than 223 million acres in state and federal grants. The settlers and business investors who acquired the parcels then formed the lines' customer base. In theory, the railroads, the government, and the settlers all benefited: the railroads got incentives to extend their lines, which in turn served farmers by transporting their agricultural goods to market and enhanced the value of the land the government still owned along the route.

But the system proved controversial. As the text under the map's title noted, "A large portion of the land thus given away is the best in the country, the railroads usually selecting the easiest routes and avoiding the mountainous and sterile districts." Even more scandalous, according to the pamphlet, was the huge amount of land thrown into limbo. The text maintained that of the 145,000,000 acres awarded them, railroads forfeited 95,000,000 by failing to construct the promised tracks. Individual states forfeited an additional 9,000,000 acres. The pamphlet implored Congress to restore these lands to the public domain and make them available to "the workers of this country" for $1.25 per acre, as specified in the Homestead Act. The author speculated that the corporations intended to sell the land to farmers for $5.00–$25.00 an acre.

The base map is credited to John Wesley Powell's *U. S. Geographical and Geological Survey of the Rocky Mountain Region.* Powell's sympathy with the message of *Map Showing How the People's Land has been Squandered upon Corporations* may have prompted him to supply its author with his survey map. Powell was outraged by the corruption rampant within the Department of the Interior, the agency that employed him and held responsibility for the public domain. The General Land Office in particular was notoriously inept and hasty, distributing 129 million acres to the railroads in less than twenty years, while granting individual homesteaders only a small fraction of the government's land.

References: The History of Our Public Lands *c. 1882; Worster 2001*

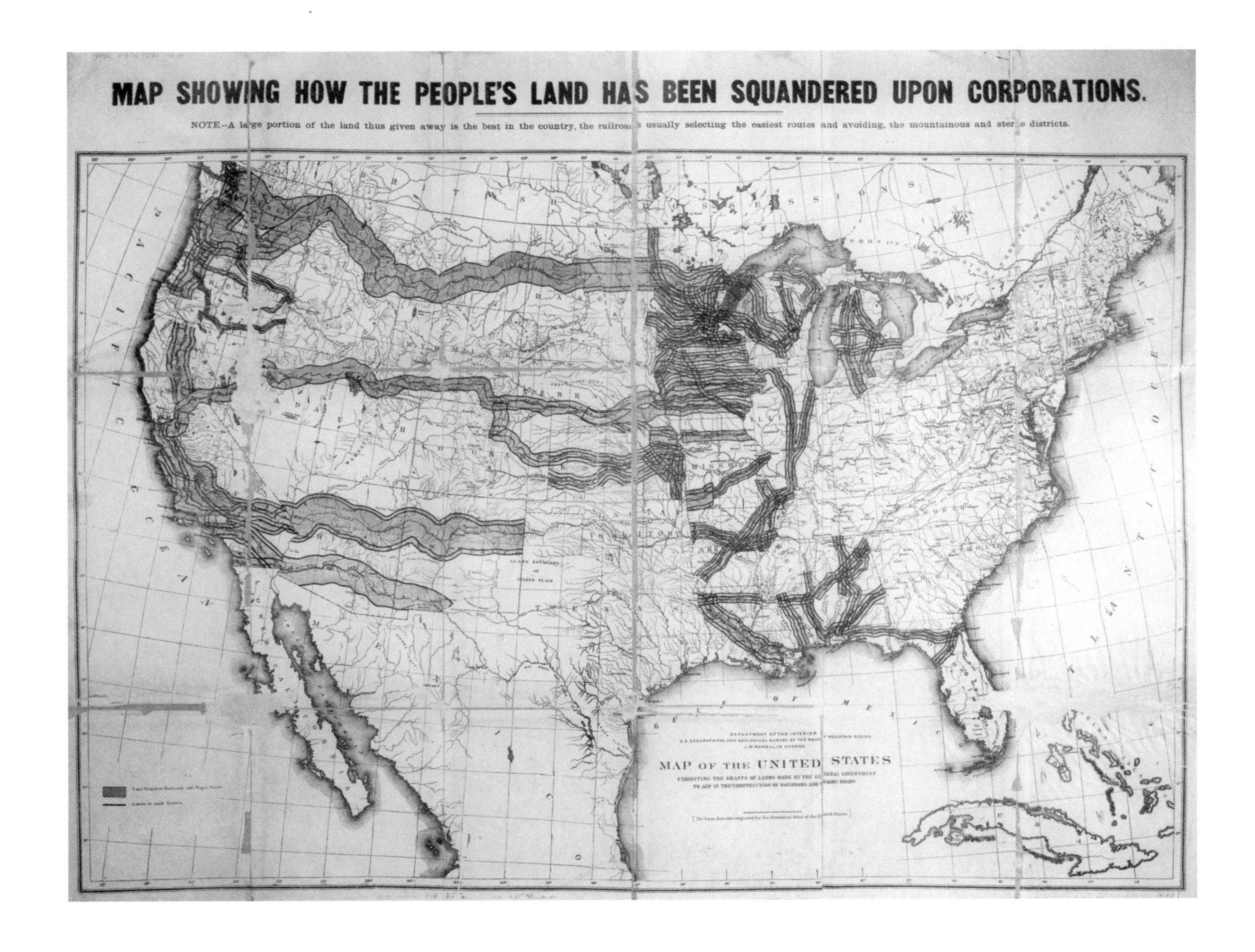

3.11

The West as a Land of Liberty

CREATOR: John F. Smith

TITLE: *Historical Geography* (Chicago: Orcutt Lithographing Co., 1888)

SIZE: 21.6 x 34 inches

LOCATION: Library of Congress, Washington, DC
[not on display]

John F. Smith's *Historical Geography* dramatizes one of the most enduring ideas about the American West: its status as a domain of freedom. A general map of the United States forms the backdrop for an allegorical diagram in which two trees competing to dominate the land represent the national struggle between liberty and slavery. According to Smith, the Pilgrims who migrated to Plymouth, Massachusetts, to escape religious persecution in Europe planted the "tree of liberty" in New England; its branches stretched westward to the Pacific coast. The Jamestown colonists imported from the Old World the "tree of slavery," which spread throughout the South until "God sent Father Abraham with his big emancipation axe to cut it down." Smith labeled the tree branches with the attributes of each culture, associating liberty with knowledge, happiness, justice, truth, and other virtues, and linking slavery to rebellion, lust, ignorance, and avarice. Smith believed that even after the abolition of slavery, the legacy of these diverging societies continued to inform contemporary politics. He noted that descendants of the Plymouth settlers belonged to the Republican Party, while the slave colony of Jamestown "still lives in the form of the Democratic Party."

Although Smith's tribute to the legacy of slavery suggests recognition of the ongoing repression of African Americans, he turned a blind eye to the subjugation of Native Americans. His graphic extension of the trajectory of liberty into the West is at odds with the government's contemporary restriction of Indians to remote reservations, and its slowness to grant them full citizenship. But the notion of Western freedom and its contribution to maintaining equality and democracy in America was widely shared and long lasting. Five years after Smith published his map, the historian Frederick Jackson Turner imparted scholarly prestige to this argument in a public address on "The Significance of the Frontier in American History." Delivered in conjunction with the 1893 World's Columbian Exposition in Chicago, Turner's "frontier thesis" soon became the most influential template for interpreting the American West. This paradigm shaped academic and popular histories of the West into the late twentieth century.

References: Ristow 1977; Schulten 2007

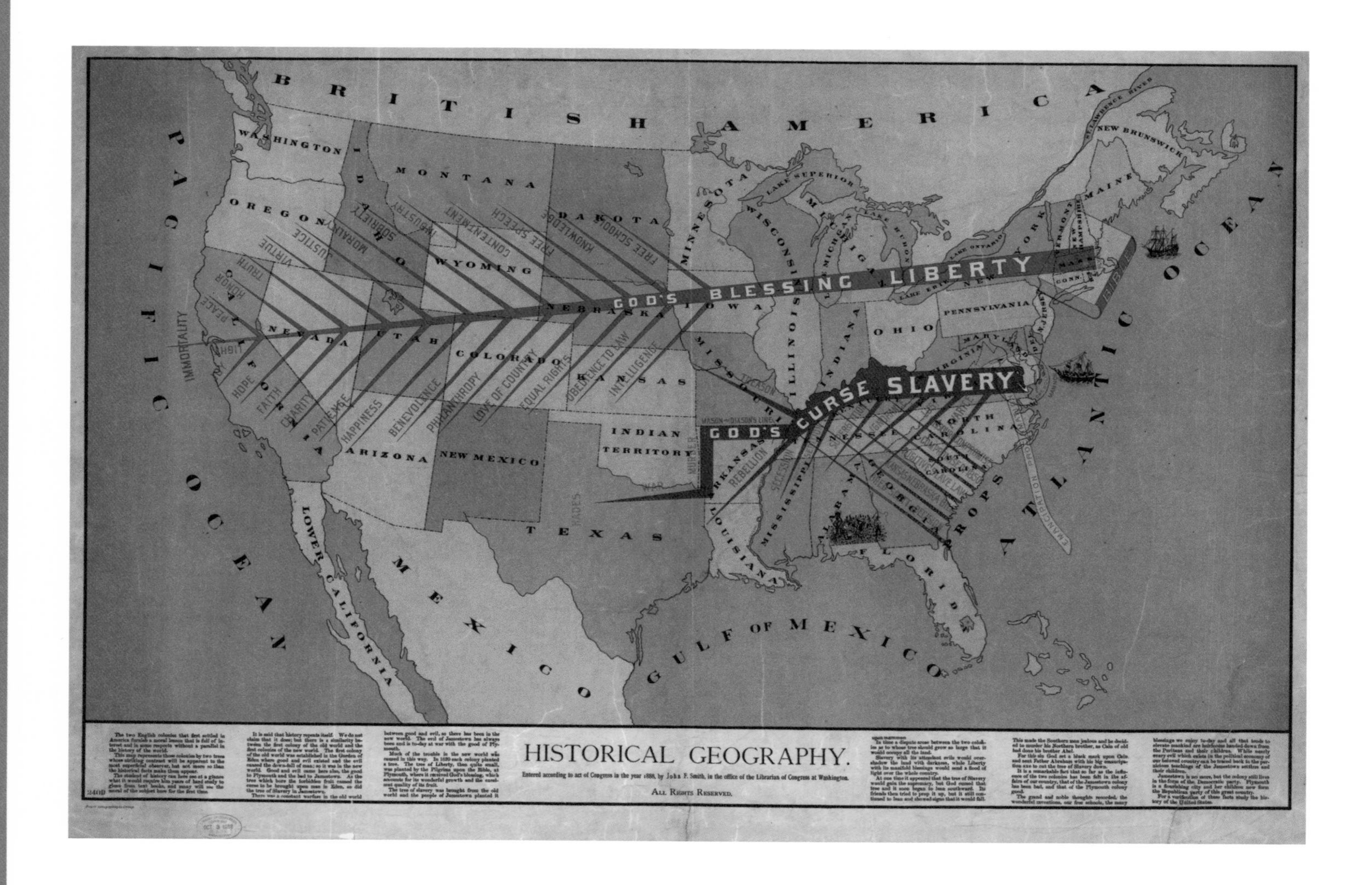

Made in Chicago: Envisioning Western Settlements

As the nation expanded westward, American consumers grew eager for information about new settlements and enterprises. Chicago mapmakers met this demand, chronicling the establishment of farms and villages, businesses and towns across the West. These cartographers often had a more direct experience of the land, a sharper sense of the pulse of the map market, and a deeper stake in the region than eastern firms. The production of local maps evolved in tandem with popular histories and business directories of cities, counties, and states. Chicago cartographers lavished special attention on their home city, but were quick to claim the entire Midwest, and parts of the Far West, as their own.

3.12

Remembering Chicago in 1812

CREATOR: Juliette Augusta Magill Kinzie

TITLE: *Chicago in 1812*, in Juliette A[ugusta] Magill Kinzie, *Narrative of the Massacre at Chicago* (Chicago: Ellis & Fergus, 1844)

SIZE: 4.13 x 7.5 inches

LOCATION: Newberry vault Case Ruggles 209

CREDIT: Gift of Rudy Lamont Ruggles

Juliette Augusta Magill Kinzie
CREDIT: Chicago History Museum

Chicago in 1812, the frontispiece in Juliette Augusta Magill Kinzie's (1806–1870) *Narrative of the Massacre at Chicago*, was the first map printed in Chicago. Kinzie drew the map to illustrate the geography of an event that she helped establish as the city's founding story. The sketch shows the extent of Chicago at the time of the War of 1812. The inhabitants clustered near Fort Dearborn at the mouth of the Chicago River included American soldiers, Indians, Indian agents, and French and British traders. A lone farm, Lee's Place, appeared down the south branch of the river. Across the river from the fort, the author marked the home of her father-in-law. A fur trader and a supplier to the fort, John Kinzie was one of the area's earliest white residents.

A month after the war broke out, William Hull, the American general in command of the northwest army at Fort Detroit, surrendered the Michigan Territory to the British and ordered Captain Nathan Heald to evacuate Fort Dearborn. When Captain William Wells and a group of Miami Indians arrived from Fort Wayne to escort the evacuees, Captain Heald destroyed the fort's supply of liquor and munitions and gave what remained to the local Indians. On August 14, the band headed south along the lakefront. They had traveled only a mile and a half when between 400 and 600 Indians ambushed the party, killing 15 Miamis and 52 Americans, and taking 41 prisoners, several of whom later perished. The attackers burned the fort the following day. The incident became popularly known as the "Fort Dearborn Massacre."

Juliette Kinzie penned her history from the perspective of her family. She had settled in Chicago with her husband, John Harrison Kinzie, in 1834—more than twenty years after the event. She based her account and map on the oral narratives of her mother-in-law, Eleanor Kinzie, and sister-in-law, Margaret Helm. Juliette Kinzie aimed to enshrine her father-in-law as the heroic "Father of Chicago." She pointed out that he had advised against the evacuation of the fort, implying that the "massacre" could have been averted if his recommendation had been heeded. Other historians maintain that John Kinzie was merely trying to protect his business interests, which depended on the fort.

References: Grossman, Keating, Reiff 2004; Holland 2005; Williams 1980

3.13

Boosting Davenport, Iowa

CREATORS: James T. Hogane (surveyor) Henry Lambach (civil engineer), Edward Mendel (lithographer, publisher)

TITLE: *Map of the City of Davenport and its Suburbs, Scott County, Iowa* (Chicago: Edward Mendel, 1857)

SIZE: 41.6 x 51.2 inches

LOCATION: Newberry Map 6F G4154 D4 1857 H6

Through its monumental scale and elaborate embellishment, the *Map of the City of Davenport and its Suburbs* expressed the tremendous personal, civic, and regional pride of local residents. The map commemorated a momentous time in this frontier city's history, appearing just a year after completion of the nation's first railroad bridge across the Mississippi River. The span connected the Rock Island Railroad on the east to the Mississippi and Missouri Railroad on the west, at a point where the river runs east-west rather than north-south. Davenport's location at an easily navigated stretch of the Mississippi, along with its proximity to coal fields in Illinois and Iowa, contributed to its selection as the site for the bridge. By making the city a link between east and west, the bridge spurred population growth, economic prosperity, and public improvements. More symbolically, the connection enabled Davenporters to see themselves as an integral part of the American epic of westward expansion.

The map encapsulated Davenport's past and present. The city's original sixty-two blocks, platted at its founding in 1839, appear at the center in pink. Each subsequent addition is identified by a different color. The river, which formed a semi-circle around the original town plan, dictated Davenport's shape. The detailed rendering of the Mississippi, including the trajectory of currents, indicated its ongoing importance. The map also marked notable public places, such as schools, a hospital, churches, and cemeteries.

Like the county wall maps of the period, the *Map of the City of Davenport* identifies all of the local property owners and enumerates their acreage. The mapmakers also adopted from county maps the practice of allowing individuals and businesses to purchase space along the border, where they could add vignettes to enhance their representation. The frame intersperses personal advertisements, such as the residence of Antoine Le Claire, one of the city's founders, with notable public places, including Iowa College and the Mississippi Bridge. In contrast to the production of most county maps by itinerant makers, local residents typically drew city maps—in this case Davenport surveyor James T. Hogane (c. 1835–?) and civil engineer Henry Lambach (1815–1899). Chicago lithographer Edward Mendel (1827–1884) published the sheet, making the map a thoroughly Midwestern product.

References: Burrows 1888; Ohl 1980; Pfieffer 2004; Rees 1854; Wilkie 1858

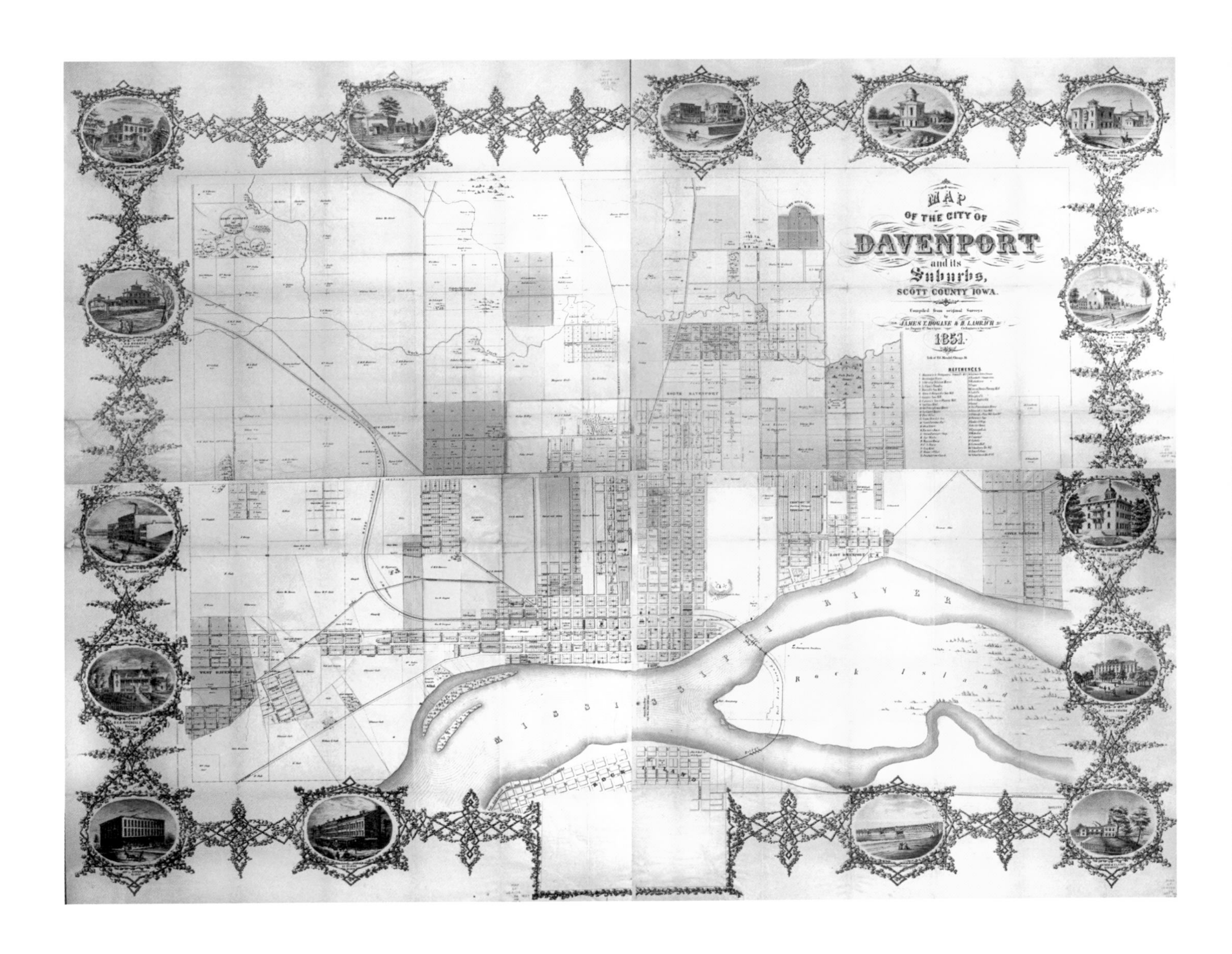

3.14

Local Pride in DeKalb County

CREATORS: Moses H. Thompson and Louis H. Everts

TITLE: *Combination Atlas Map of DeKalb County* (Geneva, Ill.: Thompson & Everts, 1871)

SIZE: 16.86 x 28.5 inches (two-page spread); 14.86 x 12 inches (map); 4.63 x 5.75 inches (Glidden residence)

LOCATION: Newberry oversize G1408 D4 T47 1871 (catalog reproduction); The MacLean Collection (exhibit)

Enterprising publishers in the Chicago area exploited, with great flair, one of the nineteenth century's most novel cartographic products: the illustrated county atlas. Thompson & Everts' *Combination Atlas Map of DeKalb County* typified the genre, capitalizing on local pride by featuring detailed maps of each township, along with historical essays, engraved views of landmarks and scenery, population statistics, and business listings. These features defined the Midwestern style of county atlas, which celebrated the region's rapid development and democratic ideals, epitomized by the abundance and prosperity of family-owned farms. The atlases' showy designs made them seem lavish, but in fact they were mass-market goods, inexpensively produced following a standard formula. Yet their appeal—and profit—turned on personal touches. Agents went door to door to sell the books in advance. For an additional fee, subscribers could have their own portraits and biographies, as well as views of their homes, farms and businesses, inserted into the volume.

Moses H. Thompson (1833–1911) and Louis H. Everts (1836–1924) were both natives of Kane County, Illinois. Thompson got his start in the trade as a railroad surveyor in the late 1850s before turning to the production of county wall maps, working with his younger brother Thomas. After the Civil War, he formed a partnership with Everts and began making atlases. From their headquarters in the Worsley block in Geneva, Illinois, Thompson & Everts produced atlases of counties in northern Illinois, southern Wisconsin, Michigan, and Ohio. The Everts firm employed a Philadelphia engraver to print their atlases, and eventually relocated there.

The entrepreneurship of the atlas makers mirrored that of their subscribers. The paid inserts in the DeKalb atlas included a view of the farm owned by Joseph Farwell Glidden, who revolutionized the agricultural settlement of the West with his development of the modern barbed-wire fence. Glidden's fences had great practical value, giving farmers an inexpensive way to protect crops from grazing animals. They also shared the symbolic function of the atlases' landownership maps, visually asserting the transformation of open land into neatly demarcated parcels of private property.

References: Conzen 1984b; Conzen 1997

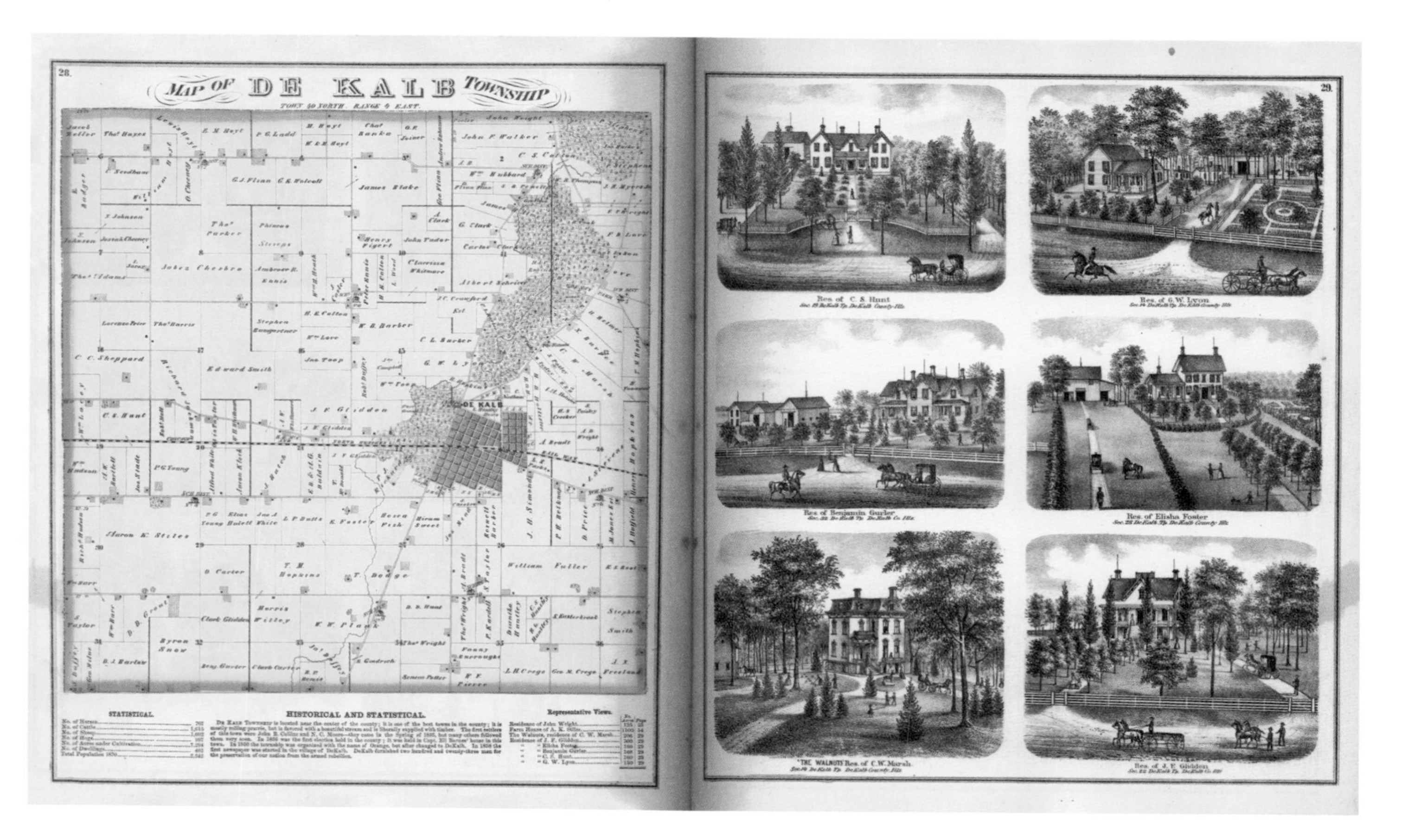

3.15, 3.16

Celebrating Minnesota

[3.15] CREATOR: Alfred Theodore Andreas

TITLE: *An Illustrated Historical Atlas of the State of Minnesota* (Chicago: A. T. Andreas, 1874)

SIZE: 18.4 x 14.8 inches

LOCATION: Newberry Case oversize G1425 A5 1874

[3.16] CREATORS: George H. Ellsbury (artist and copyright holder), Vernon Green (copyright holder), Hoffman (lithographer), Charles Shober (printer)

TITLE: *St. Paul, Minn.* from Alfred Theodore Andreas, *An Illustrated Historical Atlas of the State of Minnesota* (Chicago: A. T. Andreas, 1874)

SIZE: 15.2 x 30 inches

LOCATION: Newberry Map 4F G4144 M5A3 1894 E4 (2 of 3)

Alfred Theodore Andreas

Memorials of Deceased Companions... Military Order of the Loyal Legion of the United States, vol. 1, p. 493.

CREDIT: Newberry F 8346 334

Alfred Theodore Andreas (1839–1900) began his path-breaking career in cartographic publishing in 1867, when Louis H. Everts (see Item 3.14), a Civil War Army colleague, recruited him as a subscription agent for Everts's county map business. After Andreas hit upon the idea that packaging the same information in an atlas would offer customers a more convenient product with more space for lucrative inserts, he started his own business. Eyeing bigger opportunities, he moved to Chicago in 1873. Andreas exemplified a major shift in late nineteenth-century commercial cartography, as control of the content and visual style of maps shifted from surveyors and engravers to business managers and marketers.

When Andreas sensed that the county atlas market was nearly tapped out, he pioneered a related product: the state atlas. With this new venture, Andreas could offer the most appealing features of the county atlas to a larger audience. He targeted Minnesota for his first volume because it was ripe for mapping—only a few county atlases had been produced for the state. To orchestrate the expansive project, he established an office in St. Paul and hired Thomas H. Thompson to supervise the fieldwork of a staff that swelled to 108 members. To promote the atlas Andreas contracted with newspapers throughout the state to run his press releases. His canvassers secured 12,000 subscriptions, one from every seven households in the state. Andreas' most ambitious production to date, the Minnesota atlas owed its impressiveness to the sheer volume of text and images. Its 400 pages were filled with vividly chromolithographed maps of each county, a variety of views, and a portrait gallery of proud subscribers.

To enhance the volume's grandeur, Andreas incorporated elaborate fold-outs containing color lithographs of St. Paul, Minneapolis, and Winona. These images charted the progress of urbanization. In the St. Paul view, artist George H. Ellsbury contrasted the bucolic foreground landscape, populated by a few cows and horse-drawn carriages, with the bustling city across the river, where smokestacks and church spires pierce the skyline. The juxtaposition implied that residents of St. Paul enjoyed the best of both worlds. By including these townscapes, Andreas catered to the well-established fashion for panoramic views. Like the county atlases, the views made provincial places seem important by representing them in a distinguished mode of artistic and cartographic imagery. Most of the pictures in this tradition show cities from a bird's eye perspective. Ellsbury's vantage point was closer to the ground, making his scene seem more verifiable and accessible. Ellsbury was probably from Minnesota, as most of his surviving views picture locations in the state. Andreas may have learned of his work through Chicagoan Charles H. Shober, who printed Ellsbury's earlier views as well as the Minnesota atlas.

References: Conzen 1984a; Reps 1984

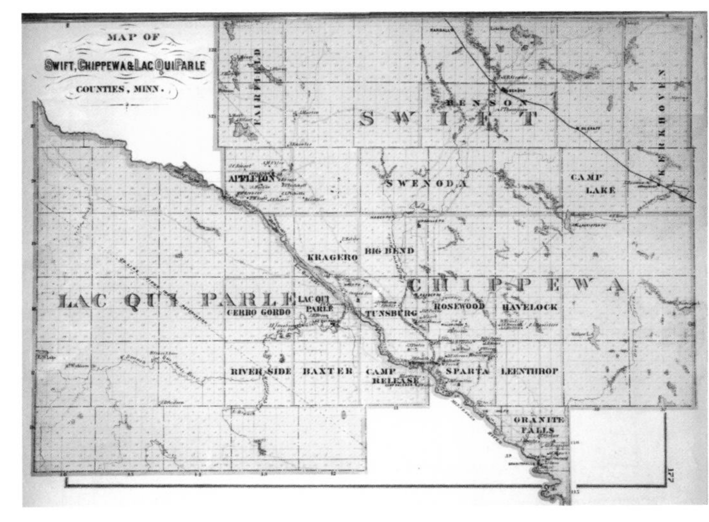

[3.15]

[3.16]

[3.15]

SECTION FOUR: MAPS FOR BUSINESS

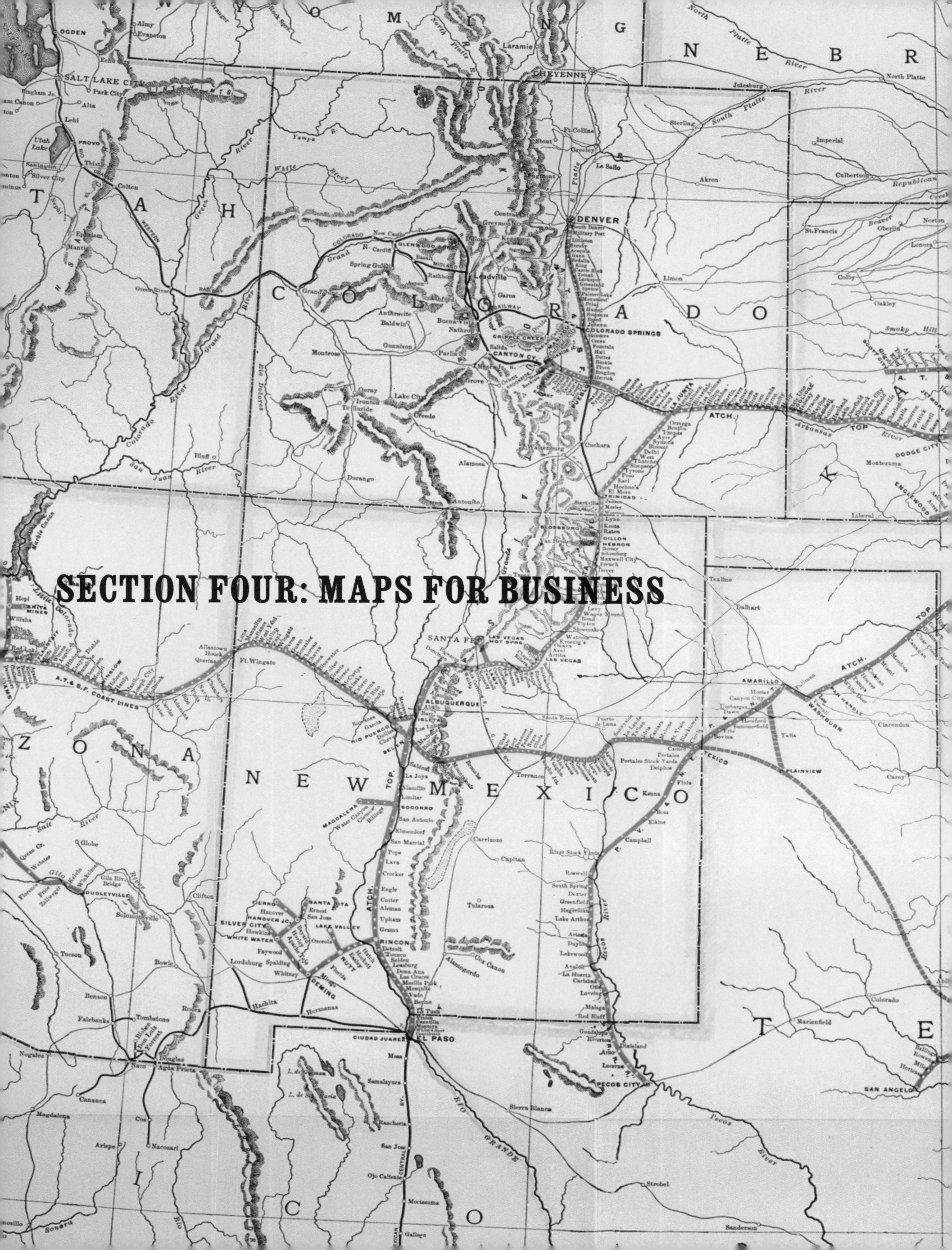

Although commercial cartographers made the general-interest maps discussed in the previous section with profits in mind, mapmakers developed distinctive genres to further specific business interests. Maps expedited the process of building transportation networks, surveying and selling land, searching for minerals, and creating tourism. By advancing Western industries and connecting them to national and international markets, maps proved crucial to the economic development of the region. Cartography itself grew into a major industry in the West, with Chicago surpassing Eastern cities as the center of the trade by the end of the nineteenth century.

Railroads Open Up the West

The explosive development of the West in the decades following the Civil War could not have happened without the railroads. With the completion of the transcontinental line in 1869, railroads united the nation, practically and symbolically. Recognizing that their success depended on the prosperity of the region, Western railroads promoted the locations they served as desirable places to live, visit, and conduct business. The companies printed and distributed an abundance of specialized maps for their investors, employees, and customers. Railroad maps pictured the links between distant parts of the country and, as operational and promotional tools, facilitated its economic integration.

Railroad companies also included handsomely-designed maps in their annual reports to chart their growth and plans for expansion. Like the reports themselves, the maps were intended to celebrate corporate achievements and ambitions for current and prospective investors. Annual report maps from the turn of the century traced the consolidation of smaller lines into regional transportation systems in the 1880s and 1890s. Led by some of the era's wealthiest and most aggressive businessmen, these consolidated railroads epitomized the monopoly capitalism of Gilded Age America.

Chicago was, literally and figuratively, at the center of these developments. By the mid-nineteenth century, the city had become the transportation gateway between the East and West. Chicago was the connecting point for two different kinds of rail networks: the high-volume trunk lines running in from the East and the fan of low-volume lines that spread out across the West. The city's position as the nation's busiest railroad hub in turn encouraged local cartographers to specialize in railroad maps. By 1875, Chicago was the leading publisher of railroad cartography in the country.

4.1

The Northern Route to the Pacific

CREATORS: Northern Pacific Railway Company, Poole Bros. (printer)

TITLE: *Great Northern Railway Map Showing Northern Pacific Railway System, Great Northern Railway System, Chicago, Burlington & Quincy R. R. System* (Chicago: Poole Bros., [1902]) from Northern Pacific Railway Company. *Sixth Annual Report of the Northern Pacific Railway Company,* 1902 (St. Paul, MN?: Northern Pacific Railway Company?, 1902?)

SIZE: 16.8 x 23.8 inches

LOCATION: Newberry Map 4F G4051 P3 1902 P6

The map included in the Northern Pacific Railway's (NPRR) 1902 report illustrates the company's amalgamation of three major lines into a system that dominated the Northwestern United States. By showing each added line in a different color, the map traced the history of the system's growth.

Chartered by Congress in 1864, the Northern Pacific gained momentum when Jay Cooke invested in the company in 1870. The railroad followed Cooke into bankruptcy in 1875, but recovered quickly, prompting Henry Villard to purchase the line in 1879. The Northern Pacific benefited from the federal government's generous land grants to subsidize the construction of railroads, receiving a total acreage that exceeded the size of Nebraska. This enabled the railroad to build the long, direct lines shown in red, encouraging the establishment of settlements along the way.

James J. Hill (1838–1916) founded the northernmost line in the system, the Great Northern Railroad, in 1889. Hill had acquired and expanded an ailing Minnesota line that had been chartered in 1856. In the absence of government subsidies, he financed the road by developing profitable industries along the route. After Hill's banker, J. P. Morgan, acquired a majority of Northern Pacific shares in 1898, the two lines operated in collusion.

In 1901 the Northern Pacific extended its reach to Chicago and Denver by acquiring the Chicago, Burlington & Quincy (CB&Q) line. As the web of blue lines on the map indicates, the CB&Q had itself aggregated many small roads. Built to carry agricultural products to local and regional hubs, these "granger" roads began in the Midwest with state charters and expanded with federal land grants in the 1850s and 1860s. Through the second half of the century, the CB&Q expanded by constructing tracks and acquiring other lines.

To produce the map, the printers, Poole Bros. of Chicago, adapted a general, wax-engraved map of the United States, cutting off the portion of the country not served by the railroad. Place names near the edge are arbitrarily sliced off, and the base map includes irrelevant information, such as topographical data and the routes of competing lines, which the NPRR doubtless would have omitted if they had produced a new map from scratch for the report. Unlike railroad maps designed for travelers, the map does not mark the stops along the route.

References: Akerman 2000; Akerman 2003; Lewty 1995; Modelski 1975; Modelski 1984; Musich 2006; Overton 1941; Villard 1904

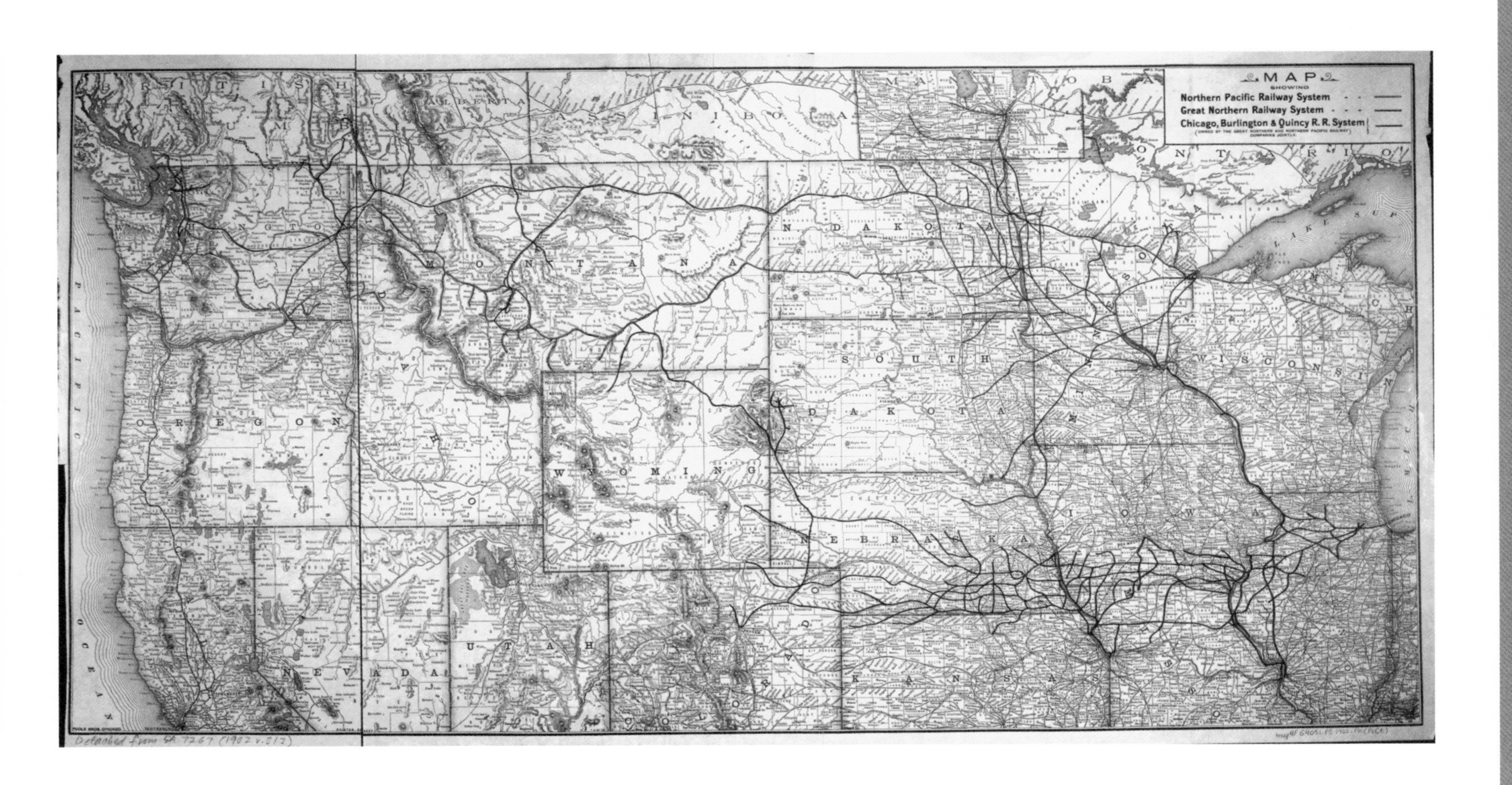

4.2

The Southern Route to the Pacific

CREATOR: Atchison, Topeka and Santa Fe Railway Company, American Bank Note Company (engraver)

TITLE: *The Atchison, Topeka and Santa Fe Railway System* (New York: American Bank Note Company, 1906), from Atchison, Topeka, and Santa Fe Railway Company, *Eleventh Annual Report of the Atchison, Topeka, & Santa Fe Railway Company ...* (New York: C. G. Burgoyne, [1906])

SIZE: 16.8 x 35.6 inches

LOCATION: Newberry Map 6F G4296 P3 1906 A4

The storied Atchison, Topeka and Santa Fe (ATSF) has dominated the southwest since the late nineteenth century. Pennsylvania-born Cyrus K. Holliday (1826–1900) founded the line in 1859 to promote the development of Topeka, a new town he helped create. Holliday financed the railroad's construction across Kansas and eastern Colorado by selling land granted it by Congress in 1863. Shortly after the line began operating in 1869, a group of Boston investors assumed financial control, bringing an infusion of capital and shrewd management.

By promoting settlement along its route, the ATSF established a customer base for its passenger and freight service. Spurs fanning out through Kansas transported agricultural commodities from farms to urban markets. After the line reached Dodge City in 1872, it hauled cattle to the city's stockyards. When it arrived at Pueblo in 1876, the railroad began carrying Colorado-mined coal to the East.

Difficulties laying track over mountainous terrain slowed the line's penetration further into the southwest. Competition for the right of way through Raton Pass in New Mexico and Royal Gorge in Colorado prompted legal battles with the Denver and Rio Grande Western Railroad in the late 1870s. After resolving its conflicts, the ATSF completed a branch line to Santa Fe in 1880, which became part of the nation's second transcontinental route when its tracks connected with the Southern Pacific Railroad at Deming, New Mexico, in 1881. But the route never met economic success, as the Southern Pacific routed freight over its own and other lines rather than through Deming.

The ATSF soon extended into California, reaching San Bernadino in 1883, Los Angeles in 1885, and San Francisco in 1900. The railroad also expanded to the south and east, opening lines to Guayams, Mexico, in 1882; to Chicago and Galveston in 1887; and into Oklahoma Territory and Indian Territory in 1899.

Tourist traffic increased as Fred Harvey (1835–1901) revolutionized hotel and food service along the ATSF. After leasing the lunch counter in the Topeka depot in 1876, Harvey built an empire that "civilized the West" with elegant accommodations, high quality food, and well-mannered waitresses famously known as "Harvey Girls." Harvey expanded his service to dining cars in 1893, and began selling Indian arts and crafts in his restaurants in the early twentieth century.

Like the Northern Pacific Railway, the ATSF adapted a general United States map for its 1906 annual report, though it carefully excluded competing lines and used the standard railroad map convention of representing stops with white circles, inserting the station names in type perpendicular to the red route line.

References: Akerman 2000; Akerman 2003; Bryant 1974; Greever 1963; Marshall 1945

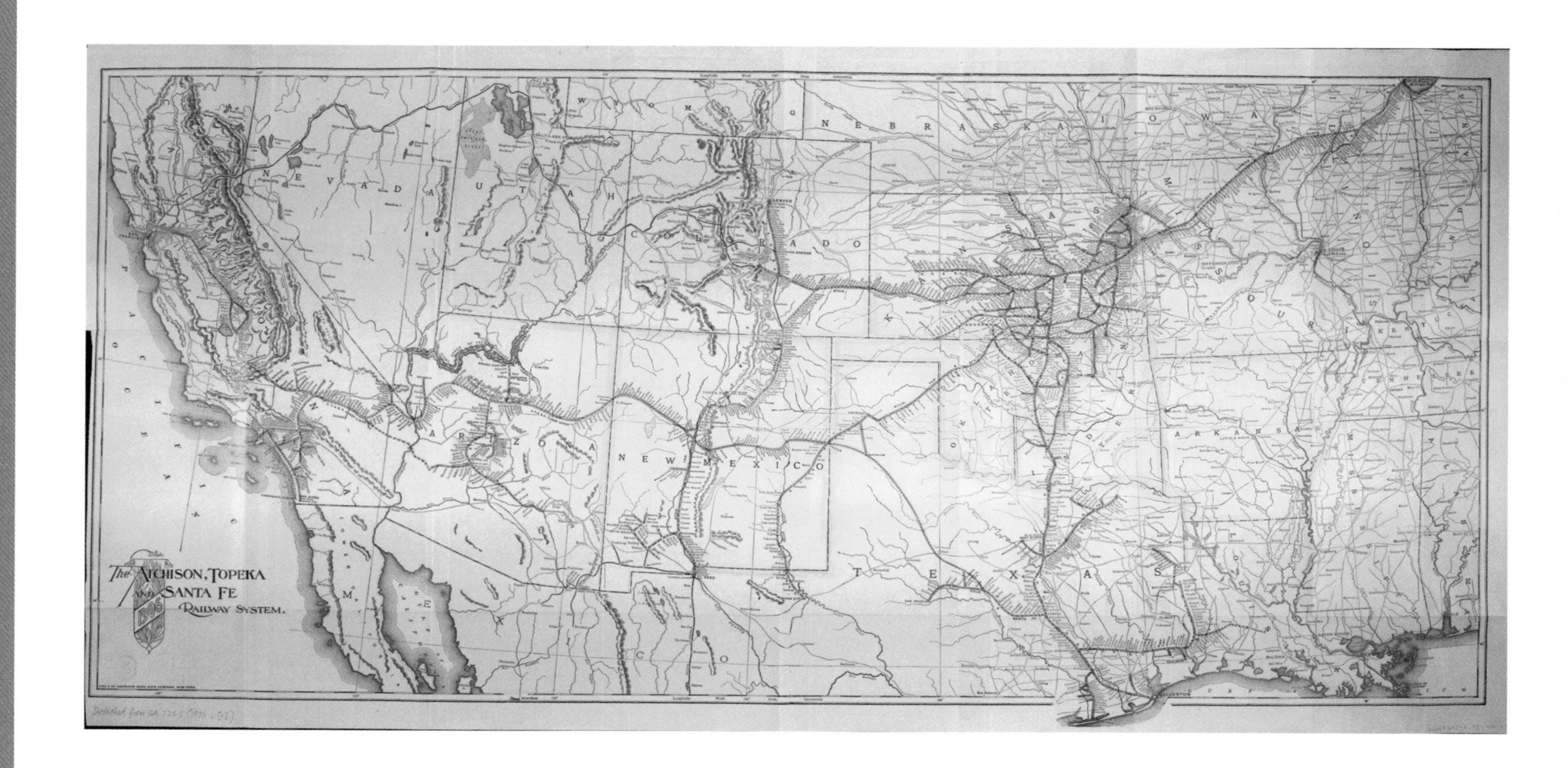

Hawking Real Estate

Throughout their history, maps and real estate have enjoyed an intimate relationship. In most traditional and early modern cultures, wealthy landowners commissioned maps of their private estates and other property holdings, which were typically passed down within families. In the United States, the increasing democratization of landownership and commercialization of property exchange in the nineteenth century prompted a boom in real estate maps. In the American West, individuals, railroad lines, mining companies, and real estate agents made and used maps to keep track of their land holdings and advertise properties for sale.

4.3

A Mysterious Midwestern Land Grant

CREATOR: Anon.

TITLE: *A Plan of Carver's Grant from the Nawdowissie Indians*, c. 1825. Pen and ink and watercolor on parchment

SIZE: 24.5 x 31.5 inches

LOCATION: Newberry vault Ayer MS map 257

CREDIT: Gift of Edward E. Ayer

A Plan of Carver's Grant from the Nawdowissie Indians illustrates an episode in a sensational saga stretching from the colonial era to the present. In 1766, the British Army dispatched Jonathan Carver (1710–1780), a Massachusetts-born soldier, to chart the rivers in what is now Wisconsin and Minnesota. He claimed to have spent the winter with a group of Dakota Sioux Indians west of the Falls of St. Anthony on the Minnesota River. After he moved to London in 1769, Carver fashioned his sojourn with the Indians into the centerpiece of his popular memoir, *Travels through the Interior Parts of North America*, published in 1778.

The land grant depicted on the map did not become public knowledge until 1781, well after Carver's death, when John Coakley Lettsom, his physician, sponsored an edition of the *Travels* that included a reproduction of the deed. Many historians suspect that Lettsom fabricated the deed, perhaps to support Carver's widow and children. In following years, numerous Carver descendants, abetted by land speculators, sought to reap profits by claiming title to the parcel and selling pieces of it.

Around 1822, Carver's heirs petitioned the U.S. Congress for ownership rights to the tract. They claimed that the Indians had given Carver the land after he negotiated peace between two warring native nations. The deed they submitted described a parcel of several hundred square miles in northwestern Wisconsin and eastern Minnesota. After an investigation, Congress denied the claim, arguing that English law disallowed land grants to individuals; that Carver himself had never mentioned the gift; and that no Indians in the area had any recollection of the deal. Despite this decision, speculators continued to secure ownership of the tract into the twentieth century.

Although no documentation of the maker or purpose of the manuscript plan survives, it was most likely made in support the 1822 petition to Congress or a subsequent claim. The view of the Falls of St. Anthony is copied from the engraving by M. A. Rooker published in Carver's *Travels*, suggesting a link to the book despite the fact that the grant does not appear in the text. The patriotic flourishes—a portrait of George Washington and an American eagle grasping the scale bar in his talons—and the division of the property into a grid of townships also make the map and the grant *seem* legitimate. But as Wisconsin was not officially surveyed until 1833, the map only approximates the government's township and range grid. The names inscribed on the large manors and squares do not correspond to individuals connected to the grant's history. Moreover, these named parcels are huge, far larger than the typical grants made to individuals through the federal system of land disposal. The enormous scale of the named tracts, along with the fertile soil and abundant minerals promised in the "Explanation," may have been designed to lure speculators or investors to the land. Although *A Plan of Carver's Grant from the Nawdowissie Indians* begs more questions than it answers, it attests to its maker's faith in the persuasive power of cartography.

References: Bosse 1986; Bourne 1906; Carver 1778; Durrie 1872; Gregory 1896; Mattocks 1867; Parker 1976; Parker 1986; Quaife 1920; U.S. House Committee on Public Lands 1825

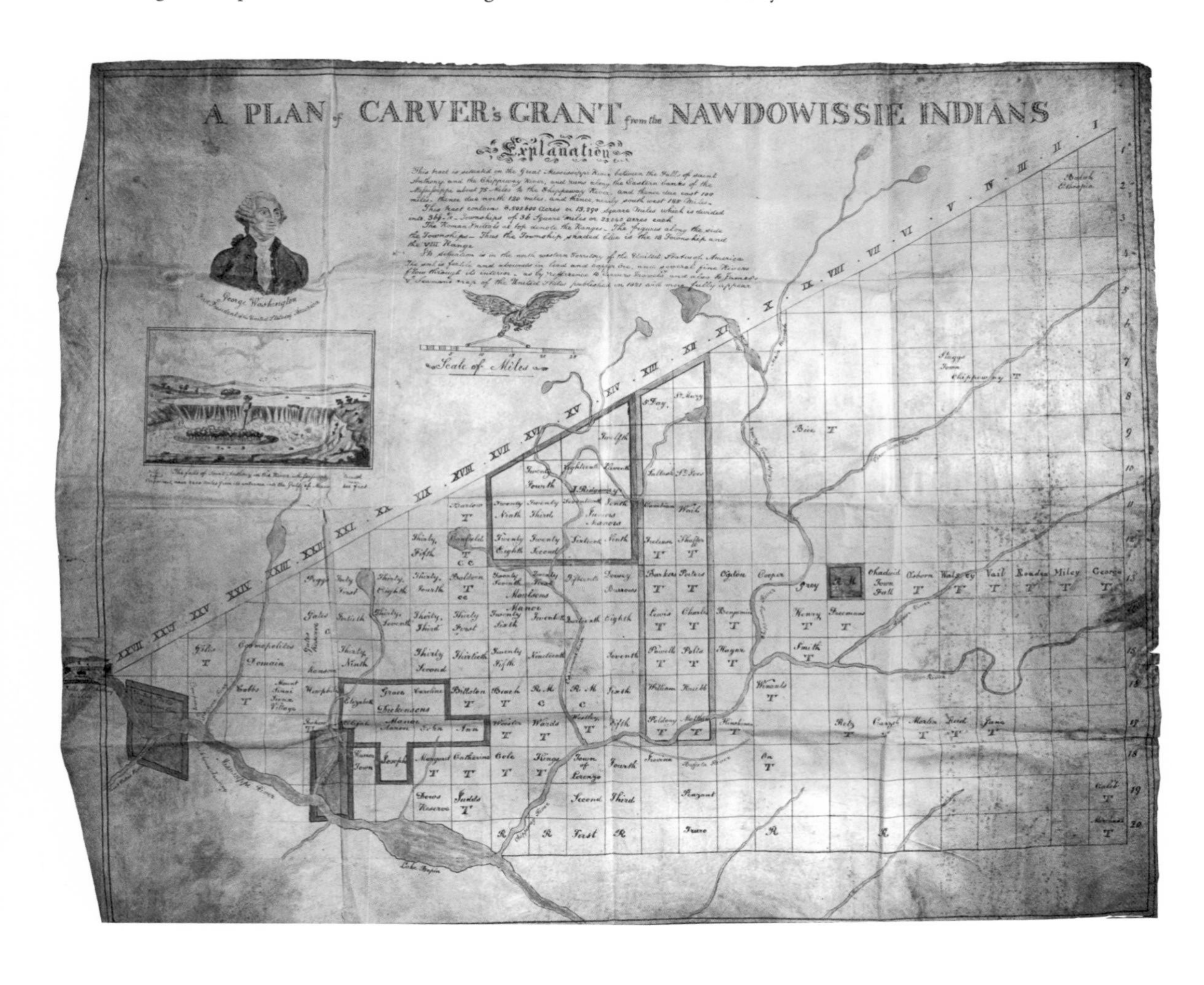

Challenging Railroad Land Grants

CREATOR: Land Department of Burlington & Missouri Railroad, H.S. Sleeper of United States General Land Office (surveyor for base map)

TITLE: Annotation of H. S. Sleeper, *Map Showing the Progress of the Public Surveys in Kansas and Nebraska* (Philadelphia: Bowen & Co., 1865); annotations and hand coloring c. 1872

SIZE: 17.2 x 18 inches

LOCATION: Newberry CB&Q map 4F G4191 G465 1865 U5

As the *Map Showing How the People's Land has been Squandered upon Corporations* makes clear (see Item 3.10), the land grant system was not without controversy. This annotated copy of the *Map Showing the Progress of the Public Surveys in Kansas and Nebraska* illuminates another of the system's complications. Burlington and Missouri Railroad officials used an 1865 General Land Office survey map as a base, atop which they diagrammed the land grants they had received in 1871 and 1872. The squares represent townships, while the colors indicate the proportion of the township included in the grant. Railroad officials probably created this document to defend their claims to the land, which were challenged in a lawsuit filed by the U.S. Attorney General. The suit charged that the sections highlighted on the map in red, blue, and green were invalid because they fell outside the customary limit of land grants extending twenty miles on either side of the route. As the inscriptions note, the Burlington and Missouri railroad had selected these more distant parcels "to make up the grant" because some of the public land within the twenty mile limit had already been sold or given to the Union Pacific or St. Joseph and Denver City Railroads. When the case reached the Supreme Court in 1878, the justices sided with the railroad. In his opinion for the court, Justice Stephen J. Field observed that much of the land had already been sold to "innocent purchasers," to whom "an incalculable amount of injustice would be done" if the grants were invalidated.

References: Akerman 2003; Musich 2006; U.S. vs. Burlington & Missouri Railroad Co. 1878; White 1991a

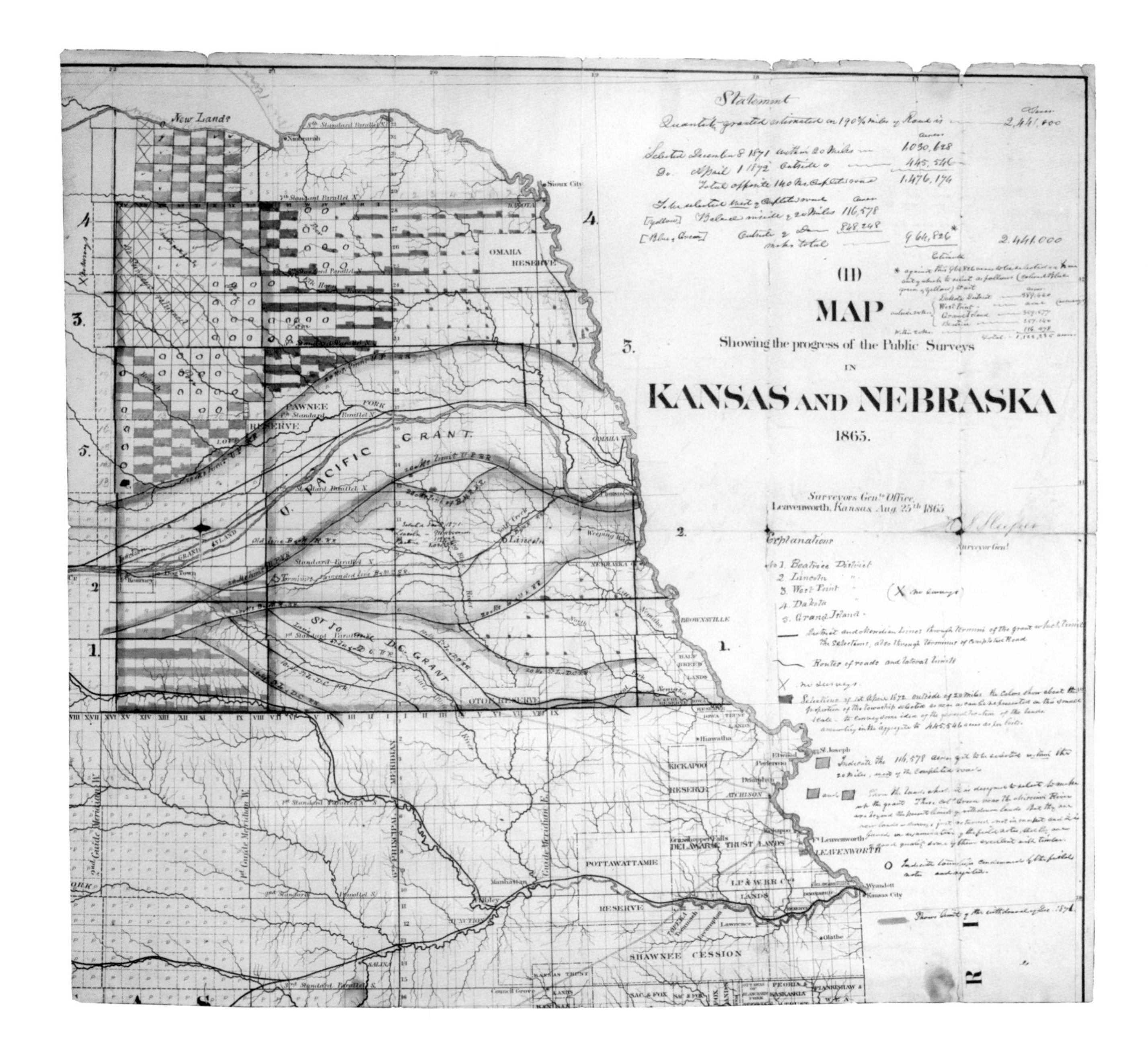

Railroads and Real Estate

CREATOR: Burlington and Missouri River R. R. Company

TITLE: *Iowa & Nebraska Land Buyers Guide* (Burlington, Iowa: Burlington and Missouri River Railroad Company, 1875)

SIZE: 21 x 29.8 inches

LOCATION: Newberry Map 4F G4151 G4 1875 F5

To carry out sales of real estate granted by the federal government (see Items 3.10 and 4.4), railroads formed land departments, which often used maps as promotional tools. George S. Harris, who took charge of the Burlington & Missouri River Railroad's (B&MRR) grants in Iowa and Nebraska in 1869, devised an aggressive campaign to lure settlers to the company's land. Many of the incentives are enumerated in the *Iowa & Nebraska Land Buyers Guide*, issued in 1875. By packaging their advertising in the form of a newspaper, the company lent an aura of objectivity to their promotions.

The inducements the railroad offered customers included the "Land Buyers' Exploring Ticket," the cost of which would be refunded upon the purchase of land. At the B&MRR's land offices in Burlington, Iowa and Lincoln, Nebraska, prospective buyers could examine plats while their baggage and families remained without charge at "Emigrant's Homes" near the stations. After customers selected property, the railroad gave them discounted rates on shipping their household goods and equipment to their new homes. The flyer also featured information of particular interest to farmers, such as the climate, soil, and rainfall patterns in each county. To further emphasize the firm's credibility, the text assured readers that the lands were offered to "actual settlers" rather than speculators and that the income from sales went directly toward the construction of the railroad. The table of prices indicates that acres were cheaper in Nebraska than in Iowa, possibly because the company had more land to sell in Nebraska. The cost of an acre ranged from \$1.00 to \$12.00 in Nebraska; \$5.00 to \$30.00 in Iowa.

Like the newspaper format, the flyer's pair of maps traded on cartography's reputation for scientific precision to disguise its bias. The smaller-scale map at the bottom shows the railroad and its land grants in a national context. The cartographer has collapsed the scale in the West, making the B&MRR's line appear at the center of the country. The major carriers from the East seem to all flow into the B&MRR's territory in what the map calls "the Gulf Stream of Migration." The map excludes the B&MRR's competitors, despite the proximity of the Union Pacific and St. Joseph and D.C. Railroads in Nebraska (see Item 4.4). The upper map is an enlargement of the lower one, offering more details about the specific land for sale.

References: Akerman 2003

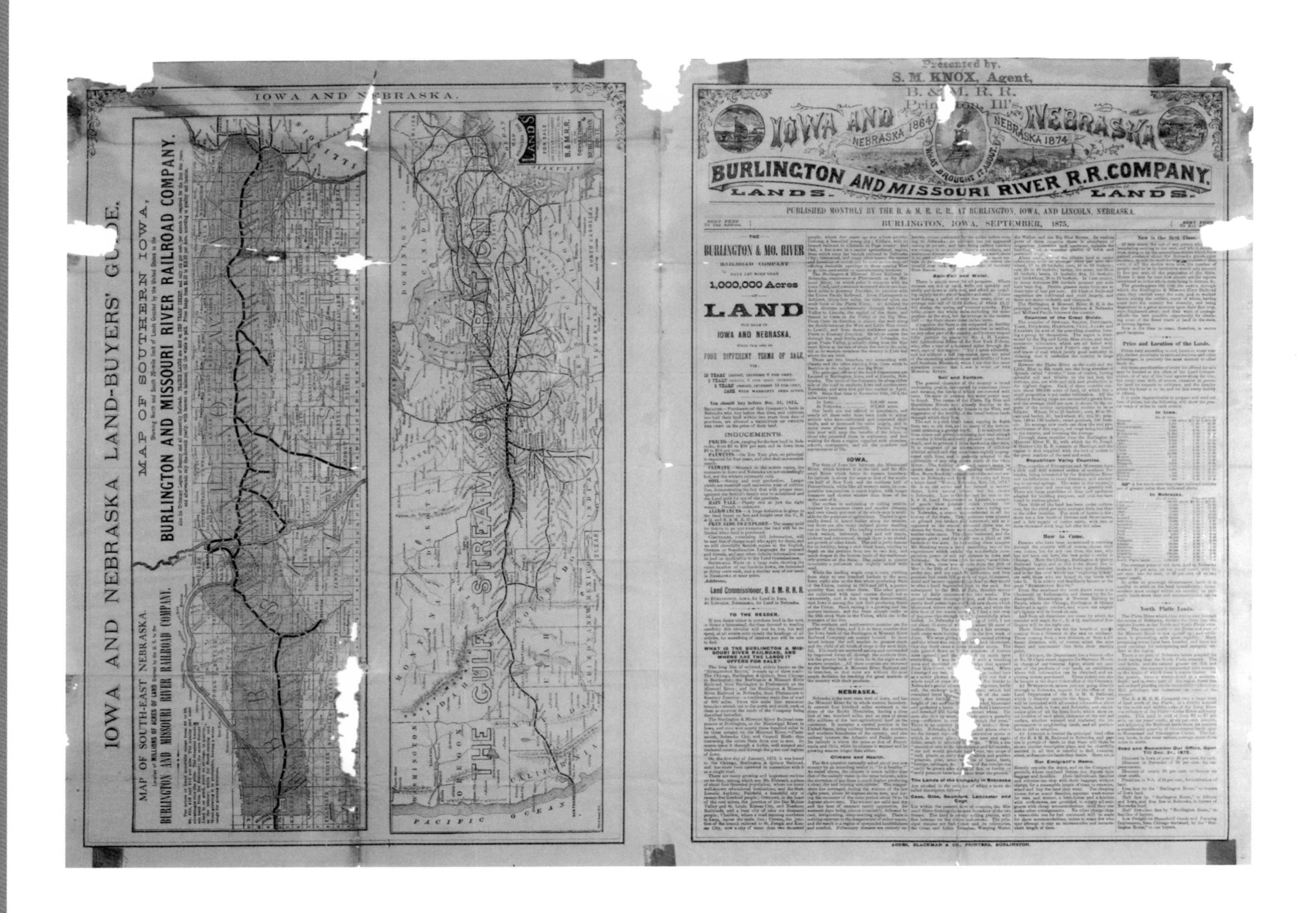

Dividing Creek and Seminole Land into Private Property

CREATORS: Ricksecker, Hockbush, and Patton (cartographers), Hall Lithographing Co. (printers)

TITLE: *Classification Map of Creek and Seminole Nations...Prepared for the Bradley Real Estate Co.* (Topeka, KS.: Hall Lithographing Co., c. 1900)

SIZE: 25.6 x 23.2 inches

LOCATION: Newberry Ayer p133 R53

CREDIT: Gift of Edward E. Ayer

The *Classification Map of Creek and Seminole Nations* narrates another episode the continuing process of opening the land reserved for Native Americans to white settlers. When the Office of Indian Affairs published the map of *Indian Territory* in 1889 (see Item 2.7), the land belonging to the Creeks and Seminoles had not yet been divided into parcels. Mounting claims from white settlers and railroads to land they regarded as not needed by the Indians led Congress to authorize a survey in 1895. The Curtis Act of 1898 dissolved the sovereignty of the Native American nations and initiated the allotment and sale of their lands.

Civil engineers Ricksecker, Hockbush, and Patton prepared the map for the Bradley Real Estate Company. Both firms had offices in Muskogee, Indian Territory. The real estate agents probably commissioned the map to distribute to homesteaders who rushed into the territory when it opened to settlers. To facilitate the newcomers' selection, the mapmakers used color codes to indicate land character. The real estate in the Seminole reserve is divided into first-, second-, and third-class parcels, while the property in the Creek reserve is classified more descriptively into Agricultural, Rocky Prairie, Hilly and Rocky, and Mountain Land. More broadly, the map summarizes aspects of the process of Indian assimilation. In addition to the land gridding, the map identifies non-native institutions already in place, including court houses, missions, and orphanages.

References: Dale 1949; Fogelson 2004; Morris, Goins, and McReynolds 1976

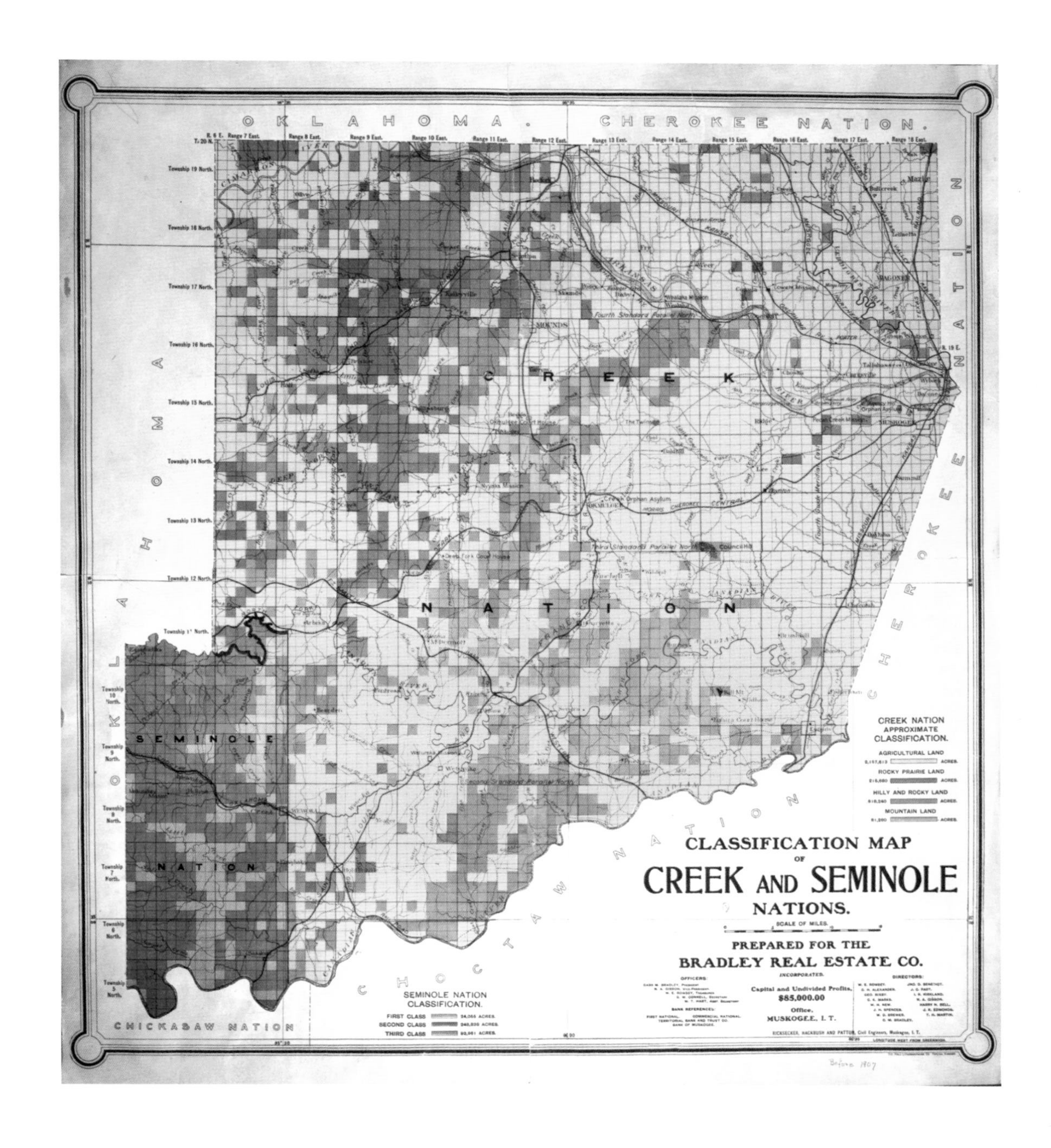

Scrambling for Mineral Wealth

The gold and silver rushes to the mid-nineteenth-century West were as rich in human drama as they were in metals. The promise of quick wealth lured hundreds of thousands of people from around the world to a region they knew little about, enticing them to endure physical dangers, hardships, and financial risks. The rushes shaped the settlement and economies of remote, arid, mountainous lands, which would have been left to the Indians were it not for the presence of precious metals.

The discovery of gold in northern California in 1848 caused the greatest excitement and set the pattern for later rushes. The first prospectors concentrated on placer mines, where erosion had already loosened the gold from its veins. The more difficult and costly vein mining followed. The intensity of California fever was sustained by the extraordinary richness of the deposits: the daily income of even an unskilled miner may have been as high as $20 in 1848. But the easy gold was quickly exhausted, dropping daily wages to $3 by the end of the 1850s. When gold was uncovered in Colorado and silver in Nevada in 1859, many miners headed over the Sierra Nevada to tap the new lodes. Smaller rushes followed gold strikes in Colorado, Idaho, and Montana in the mid-1860s and in Dakota in 1874. Rich silver finds followed at Leadville, Colorado, and Tombstone, Arizona in 1877.

Maps proved crucial at the rushes' every stage. Transportation maps guided miners to lodes, while maps of specific mineral areas charted the local topography, identifying deposits and claims, mills and towns. Technical diagrams showed the configuration of shafts and tunnels, while promotional maps helped speculators attract investors.

To the California Gold Fields

CREATORS: Ensigns & Thayer (publishers)

TITLE: *Map of the Gold Regions of California, Showing the Routes via Chagres and Panama, Cape Horn, etc.* (New York: Ensigns, Thayer and Co., 1849)

SIZE: 14.8 x 17.2 inches

LOCATION: Newberry Map 4F G4361 H2 1849 E5a

Fresh advances in communication and transportation fueled the rush to the California gold fields in 1848–49. News of the discovery of gold on John Sutter's land (near present-day Sacramento) spread quickly, aided by the expanding distribution of newspapers, broadsides, and mail. Fast clipper ships and ocean steamers carried passengers to the Pacific coast before the transcontinental railroad was completed in 1869. Would-be prospectors obtained travel advice and basic information about the area from the promotional *Map of the Gold Regions of California*.

The broadside promised beautiful scenery and abundant agricultural resources as well as gold of "superior quality." The "remarkably healthy" climate had "no winter," and San Francisco's fine harbor was well situated for maritime commerce with Asia. The journey to California was another matter. The text described in the greatest detail what may have been the most common route via the Isthmus of Panama, warning travelers that Chagres, the port on the Gulf of Mexico, was an unsightly village where the climate was unhealthy and the supply of provisions unreliable. The overland passage to the Pacific was slow, and upon reaching Panama one would encounter poor accommodations and dodgy food. The broadside recommends an alternate route across southern Mexico as "the cheapest, quickest, and safest." The all-water trip around Cape Horn was "the most acceptable as far as cost and facilities are concerned," but took the most time.

The instant demand for information sparked by the gold rush led publishers to adapt maps already on hand. New York mapmakers Ensigns & Thayer relied on their 1847 *Pictorial Map of the United States*, adding a golden patch to mark California's mining region. Based on Preuss's 1845 map of the Frémont expedition (see Item 2.2), this sheet delineated Oregon and California with reasonable accuracy.

The Newberry's copy of the *Map of the Gold Regions of California* was inserted into Thomas Jefferson Farnham's *Life, Adventures, and Travels in California* of 1849. Farnham (1804–1848), a Vermont lawyer, led an 1839 expedition to Oregon and then went on to California. He recounted his adventures in a series of popular books that promoted westward expansion. Because not all copies of Farnham's *Life, Adventures, and Travels in California* included the map, it was probably added to some copies after printing. The insert may have been a special premium offered to promote the book's sales; canvassers hawking subscriptions to the volume may also have carried the colorful sheet to stimulate sales. Some copies of the map were printed on thick paper to serve as broadsides rather than book inserts.

References: Churchill 1991; Farnham 1849; Wheat 1942; Wheat 1949

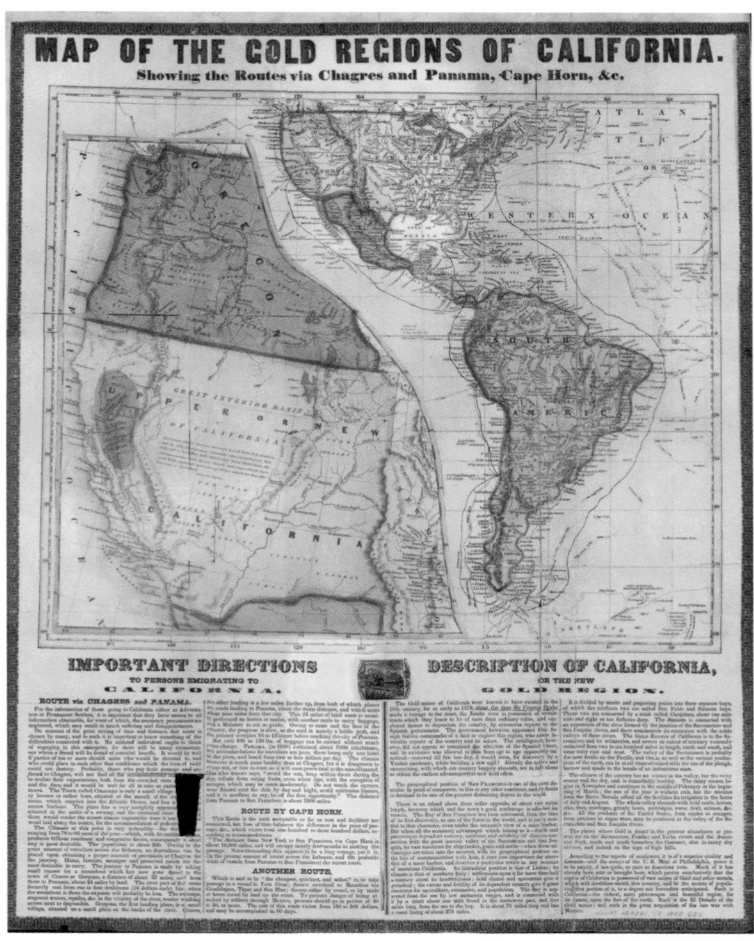

Gold Mining and Town Building

CREATORS: John J. Pratt (surveyor), Bela S. Buell (recorder of mining claims), Frederick J. Ebert (surveyor, inset map), John E. Dillingham (delineator, vignettes), W. H. Rease (lithographer and printer), with annotations by an unidentified author

TITLE: *Map of the Gold Regions in the Vicinity of Central City, Gilpin Co., Colorado Territory* (Philadelphia: W. H. Rease, 1862), with later annotations in brown ink

SIZE: 106 x 129 inches

LOCATION: Newberry vault Graff 3344

CREDIT: Gift of Everett D. Graff

After John H. Gregory discovered gold in the hills in 1859, Central City, Colorado, grew up at lightning speed. John Pratt, a surveyor, and Bela Buell (1838–after 1910), the first elected recorder of mining claims in Central City, charted this transformation in the promotional map they issued three years later. Combining a large map of the gold region, an inset map of Colorado, and 27 pictorial views, the sheet summarized the development of the Central City area. Pratt and Buell (who also aggressively bought and sold claims while serving as recorder) aimed to lure investors and settlers to the area by showing its prosperity and civility.

On the main map, Pratt inventoried topographical features pertinent to mining—identifying lodes, gulches, and ditches, and charting hills with richly drawn hachure lines. He also marked man-made incursions, including town plats, mills, toll roads, and railroads. The names of several mills just north and west of the town Enterprise—Chicago Co., Black Hawk Co., St. Louis Co., Milwaukee Co.—indicate that they were owned by Midwesterners.

The inset map in the lower right shows the entire Colorado Territory, drawn by Frederick J. Ebert (1823?– ?) under the direction of Governor William Gilpin (see Item 3.9). Although the territory had been organized only a year earlier, in 1861, the map highlights its political progress by outlining in color the boundaries of its seventeen counties and its Indian Reserve on the eastern Plains. Ebert also located the many towns and mining camps in the mountains as well as the territory's main roads. He relied on recent government maps for the topography, and included initial surveys by the General Land Office.

The ornamental border vignettes emphasize the map's promotional purpose. Recalling the format of county wall maps, the views were paid for by the businesses they picture. Pratt and Buell (or their agent) probably sold these spaces when they canvassed the area selling advance subscriptions to the map. The pictured enterprises illustrate the rising sophistication of Central City and nearby Denver and Colorado City. Several are mining related, such as A. J. and J. M. Van Deren's quartz mill, which separated gold from the soft quartz in which it was embedded. In 1859, Kansas bankers Clark, Gruber & Co. hurried to establish a private mint to coin the gold. Clark traveled to Philadelphia to purchase the equipment and had it shipped to Denver. By July 20, 1860, the mint was operating at the corner of G and McGaa Streets. Other enterprises—a brewery, stores, newspaper offices, stage offices, and a post office—served the miners and other residents drawn by the boom. While these illustrations advertised individual businesses, the map as a whole promoted the region more generally. The mapmakers filled in the unsold slots with views of sublime Colorado scenery and panoramas of Central City and Denver, assuring potential settlers that they could enjoy both nature and culture. John E. Dillingham (fl. 1860–1880), an artist from Connecticut who came to Colorado after living in Chicago for four years, sketched all of the vignettes, which he also sold separately.

Although the map would have boosted local pride when hung to decorate the walls of local homes and businesses, it also served more utilitarian purposes. An unidentified owner annotated the Newberry's copy, marking new lodes west and south of Mountain City and demarcating individual mining districts (operated by companies including the Illinois Central Railroad) in brown ink. This updated map may have been intended to inform investors about the latest gold discoveries and mining claims.

References: Bancroft 1958; Greever 1963; Mumey 1948; Perrigo, 1934; Wheat 1957–1963, vol. 5, part 1

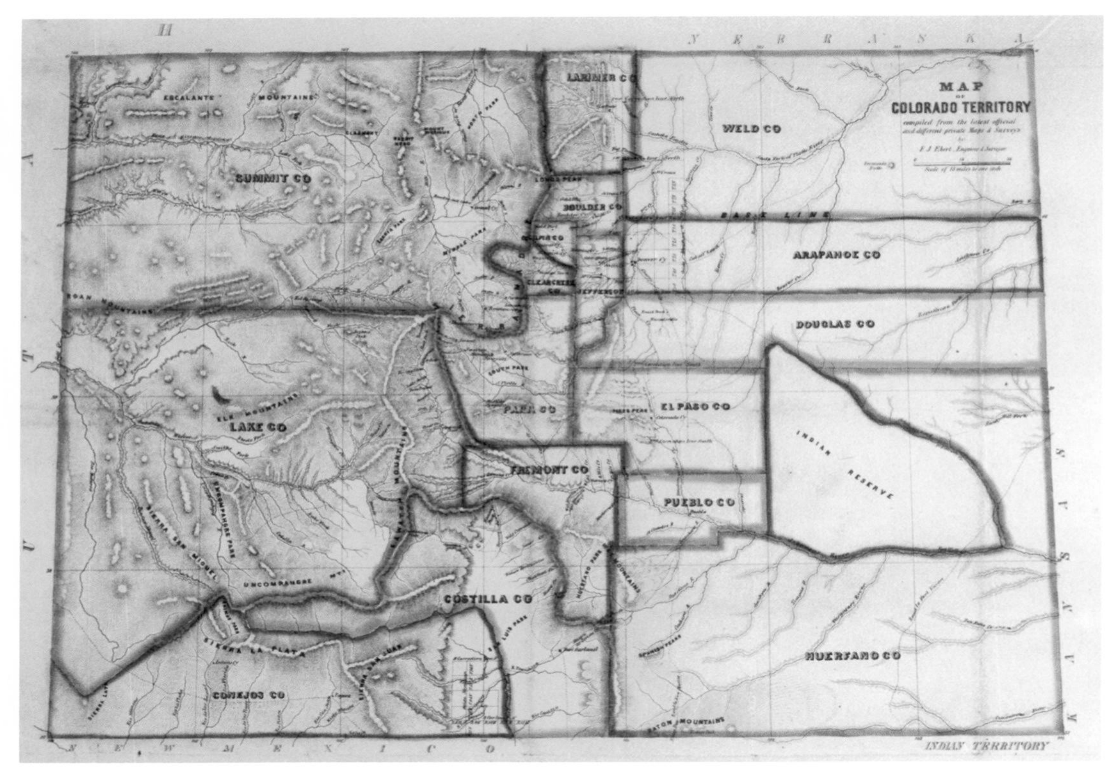

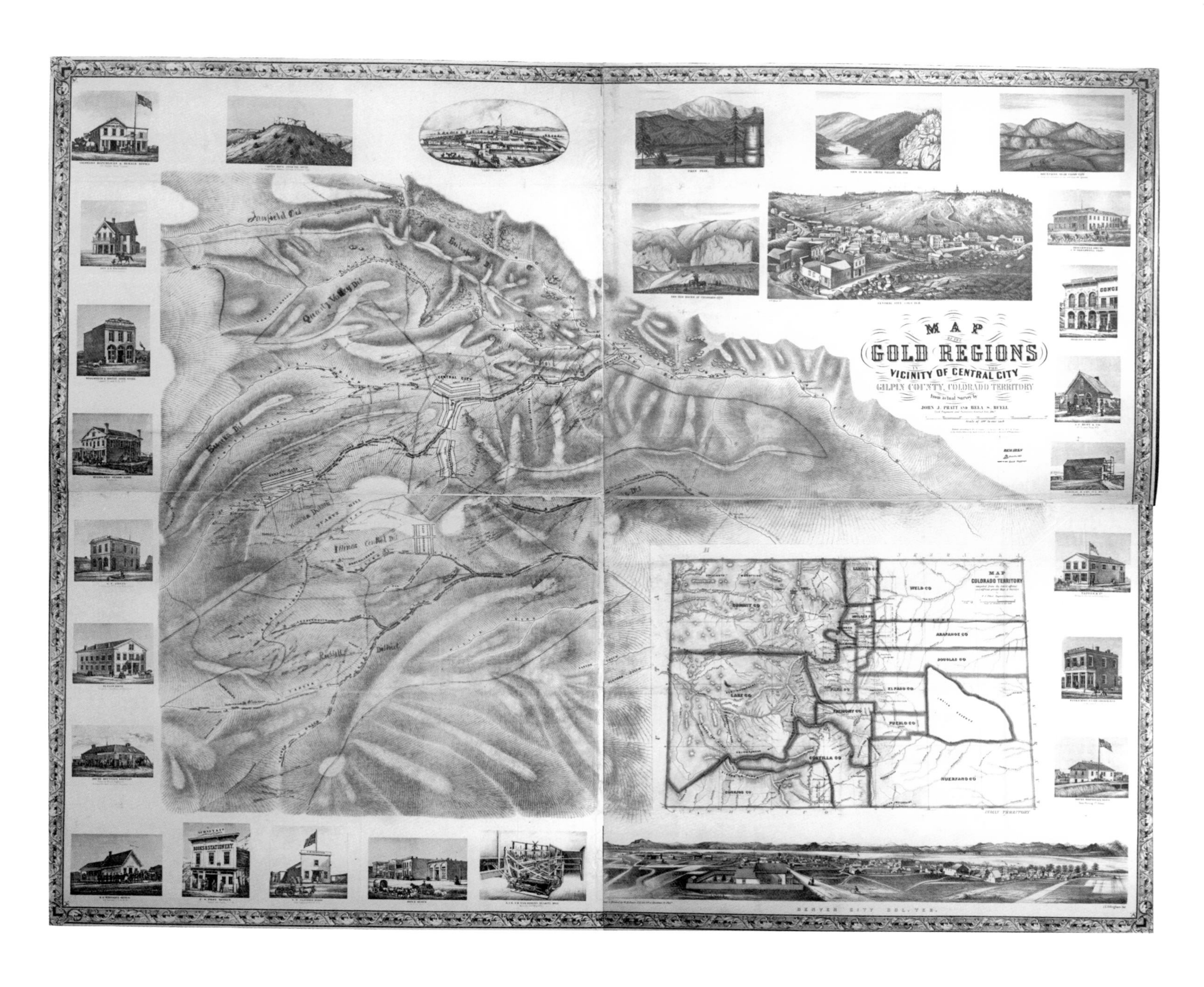

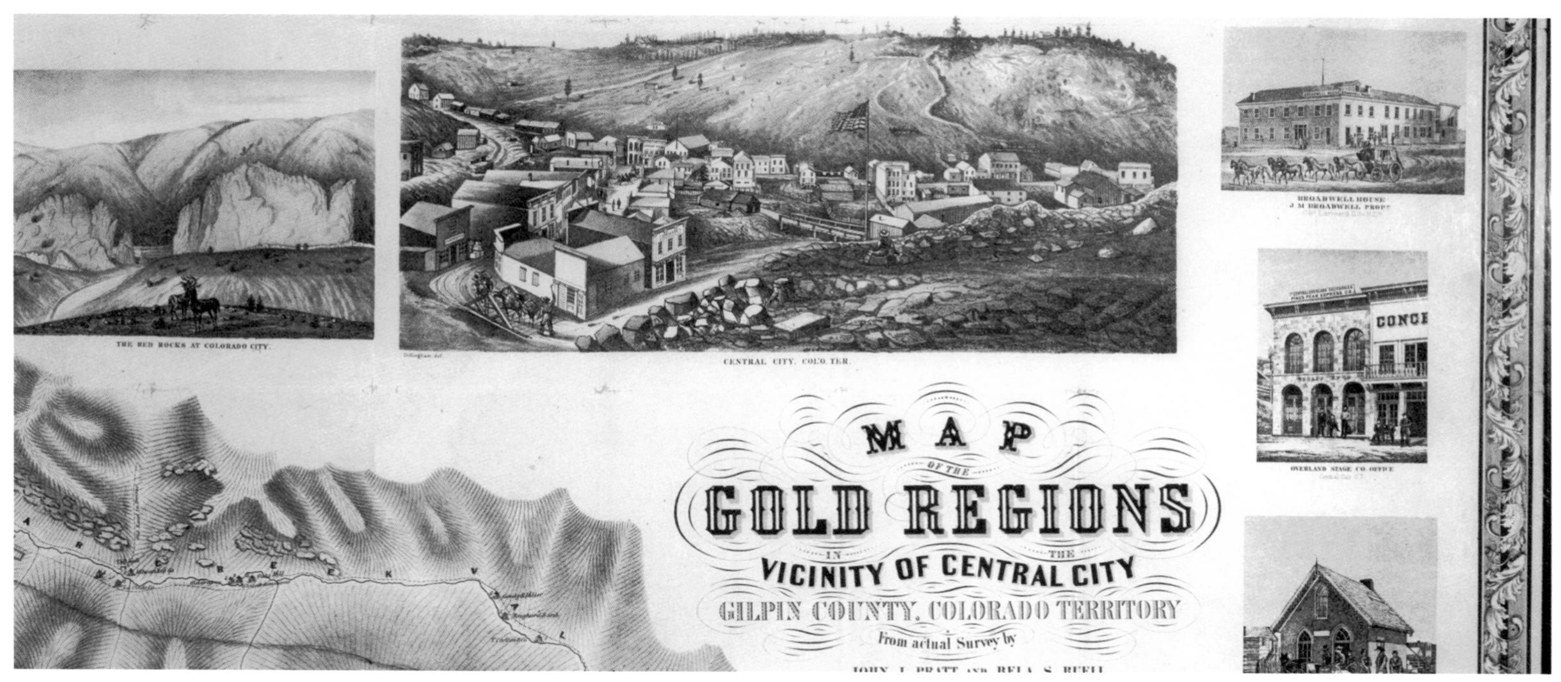
THE RED ROCKS AT COLORADO CITY.
CENTRAL CITY, COLO TER.
BROADWELL HOUSE
J M BROADWELL PROPR
OVERLAND STAGE CO. OFFICE
CONCE
MAP
OF THE
GOLD REGIONS
IN
THE
VICINITY OF CENTRAL CITY
GILPIN COUNTY, COLORADO TERRITORY
From actual Survey by

4.9

Tunnels to the Comstock Lode

CREATORS: Clarence King (survey leader), R. H. Stretch (surveyor, draftsman, compiler), I. E. James, Marlette and Hunt, T. D. Parkinson (surveyors), United States Geological Exploration of the Fortieth Parallel

TITLE: *Horizontal Map [of] Virginia Mine Workings, Comstock Lode, 1870*, in United States Geological Exploration of the Fortieth Parallel, *Atlas Accompanying Volume III on Mining Industry*, Plate 4 (New York: Julius Bien, 1870)

SIZE: 19.5 x 48 inches

LOCATION: Newberry oversize M 7583 907 1870

Clarence King
CREDIT: New York Public Library

In 1867 the U.S. Department of War appointed Clarence King, a 26-year-old civilian, to survey the land along the 40th parallel. King's project was the first of four post-Civil War government surveys of the West. His territory reached from the 105th meridian westward to the 120th, extending far enough north and south to encompass the lines of the Central and Union Pacific Railroads. King won the appointment by arguing that the mountains along the routes were not worthless territory, but "full of wealth" waiting to be unearthed.

King recorded his findings in meticulously-detailed reports, photographs, and diagrams. The first published report, *Mining Industry* (1870), was accompanied by this oversized atlas of fourteen maps and geological sections. The report and atlas lavished the most attention on the richest vein in the survey district, the Comstock Lode at Virginia City and Gold Hill, Nevada. In contributions by King and three other scientists, James D. Hague, Arnold Hague, and S. F. Emmons, the volume offers a wealth of technical information about the geology of the mining districts, the engineering of the shafts, and the equipment and chemical processes used to extract minerals.

The *Horizontal Map [of] Virginia Mine Workings* is one of three plates in the atlas tracing the complex configuration of underground shafts. The map collapses depth, representing all of the mines as if they stretched across the surface of the ground. Thus, tunnels at different levels seem to overlap. To clarify that the mines existed on different planes, the cartographer drew tunnels at different depths in different colors, using deep gold and red for the deepest shafts, pale blue and lavender for those closer to the surface, and so forth. Small black rectangles mark the shafts connecting the mine tunnels to each other and to the surface. Other plates in the atlas present longitudinal profiles, showing the location of silver bonanzas, and cross sections specifying the depth of various types of rock. Although the dispassionate style of the essays and maps make them seem purely scientific, virtually all of the data served the economic goals of the mining companies.

References: Bancroft 1958; Bartlett 1962; Hague 1870; Trachtenberg 1989; Wright 1877

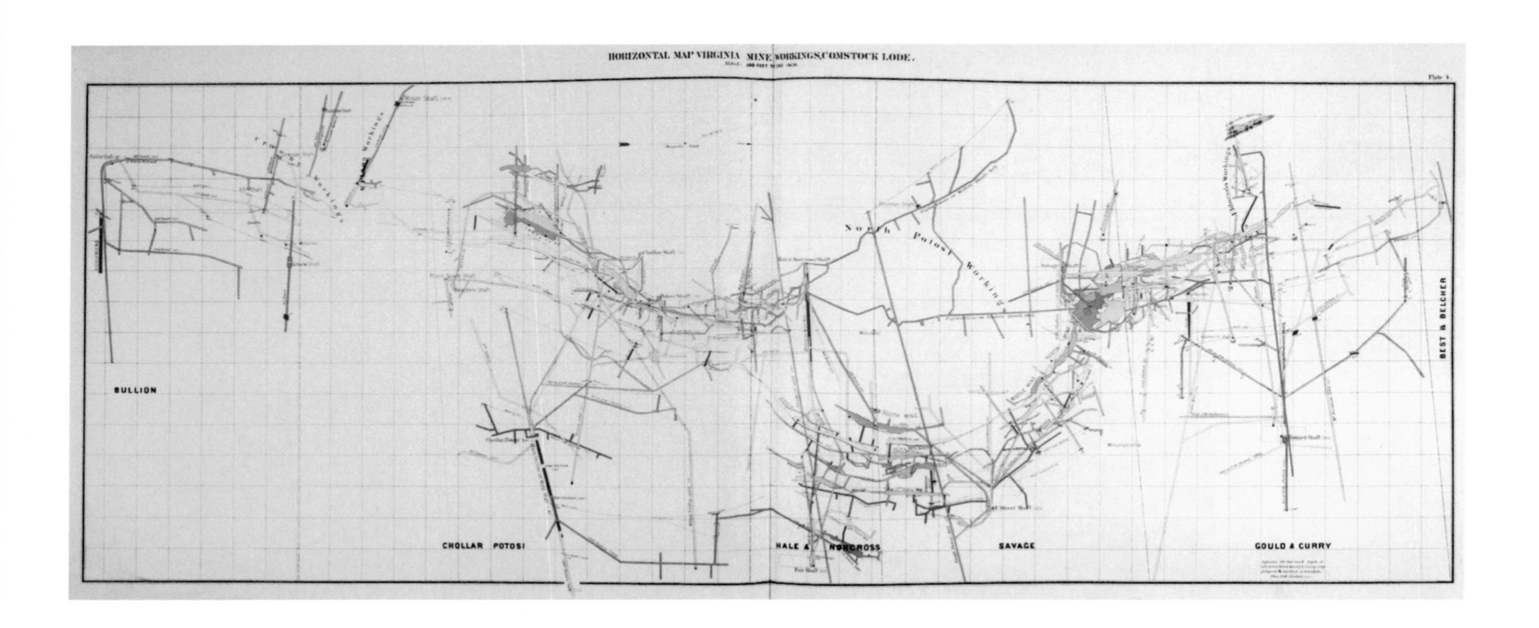

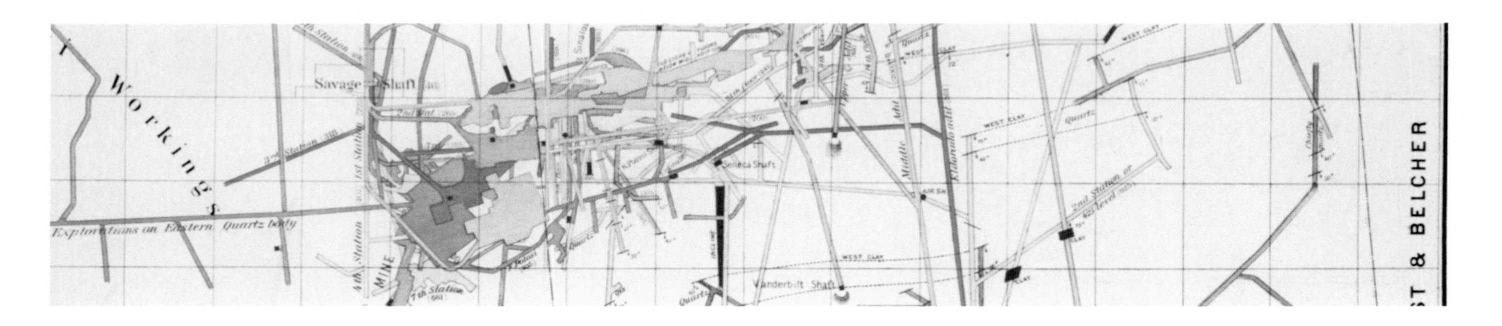

4.10

Claims to the Comstock Lode

CREATORS: T. D. Parkinson, C.E. (surveyor) and Grafton Tyler Brown (printer)

TITLE: *Map of the Comstock Lode and the Washoe Mining Claims in Storey & Lyon Counties, Nevada* (San Francisco: G. T. Brown & Co., c. 1875)

SIZE: 26.4 x 32.8 inches

LOCATION: Newberry Graff 3200

CREDIT: Gift of Everett D. Graff

Grafton Tyler Brown

Grafton Tyler Brown

CREDIT: Courtesy British Columbia Archives, Image #A-08775.

In contrast to the scientific tone of King's volumes (see Item 4.9), T. D. Parkinson focused more obviously on economics, diagramming the complicated jigsaw of claims to the Comstock Lode. Because extracting silver from veins demanded considerable skill and capital for labor and equipment, many would-be miners aimed to get rich by speculating in claims or shares in claims.

Controversies hovered over the Comstock claims from the start. In the early 1850s, prospectors in Gold Canyon, a ravine running from Mt. Davidson to the Carson River, grew irritated when their gold was polluted with "blue stuff." Two knowledgeable brothers, Ethan Allen Grosch and Hosea B. Grosch, recognized the impurity as silver. After both died suddenly, another miner, Henry T. P. Comstock, took possession of their cabin and learned about the silver ledge from the papers and ore samples they left behind. When yet another pair of miners, Peter O'Riley and Patrick McLaughlin, struck a rich part of what would become known as the Comstock Lode, Comstock and his friend Emmanuel Penrod hurried to assert ownership of the land and demand a portion of the claim. Penrod, Comstock, & Co. started the Ophir mine around which the town of Virginia City sprang up.

Big profits eluded the earliest miners in the area, falling instead to subsequent investors. When the rush began in earnest in 1860, miners staked over four thousand claims in one year within thirty miles of Virginia City—most of them worthless. Others whose claims were valuable (including O'Riley, McLaughlin, Comstock, and Penrod) knew too little about silver mining to profit from it. They expected their claims to be exhausted in a few months (as gold mines would be) and sold their stakes for a few thousand dollars. Capitalists from California snapped up the ground, and financial control shifted to absentee landlords from San Francisco. In 1862, 37 brokers formed the San Francisco Stock and Exchange Board, through which they sold mining shares on commission. The traders telegraphed their results to Virginia City, Gold Hill, and other mining towns, where local brokers posted prices on bulletin boards displayed in their windows.

Parkinson produced his *Map of the Comstock Lode* to advertise the sale of shares. Lithographed in San Francisco by G. T. Brown & Co., the map was probably aimed at California investors. Inside the cover enclosing the sheet, Parkinson included a list of the claims outlined on the map, indicating the size and number of shares available for each claim. He paid scant attention to the area's topography, concentrating instead on the geography of the claims and man-made features pertinent to the industry: toll roads, railroads, shaft openings, tunnels, and mills. The presence of a hospital and grave yards point obliquely to the dangers of mining.

The longitudinal section across the top of the sheet shows the disposition of the Comstock Lode. It was not a continuous vein of hard quartz, but was composed instead of irregular bodies of rich ore deposited at varying intervals. This configuration meant that some shafts struck silver bonanzas, while others penetrated barren rock. The discovery of the famed Big Bonanza of the Consolidated Virginia mine in 1873 created a second boom, quieting the fears expressed in King's 1870 report that the lode was nearly spent. The Big Bonanza prompted the organization of the California Company, controlled by the Consolidated Virginia, and yielded $150,000,000 in metal. It also sparked ruinous inflation. By 1875, the year Parkinson issued his map, stock prices for the Consolidated Virginia and the California mine had dropped dramatically. The map may have been an attempt to reverse the decline by attracting new investors. New methods of extraction sparked a revival in the mid-1880s, but by 1890, the area began a precipitous decline.

Although the details of Parkinson's career are obscure, Grafton Tyler Brown (1841–1918) is recognized as the earliest African-American printmaker in California. A Pennsylvania native, Brown moved west during the gold rush, seeking to escape racial prejudice and take advantage of the economic boom. After running his own lithography business in San Francisco, Brown pursued a career as a landscape painter, specializing in Western scenery.

References: Bancroft 1890; Greever 1963; Lamar 1998; Wright 1877

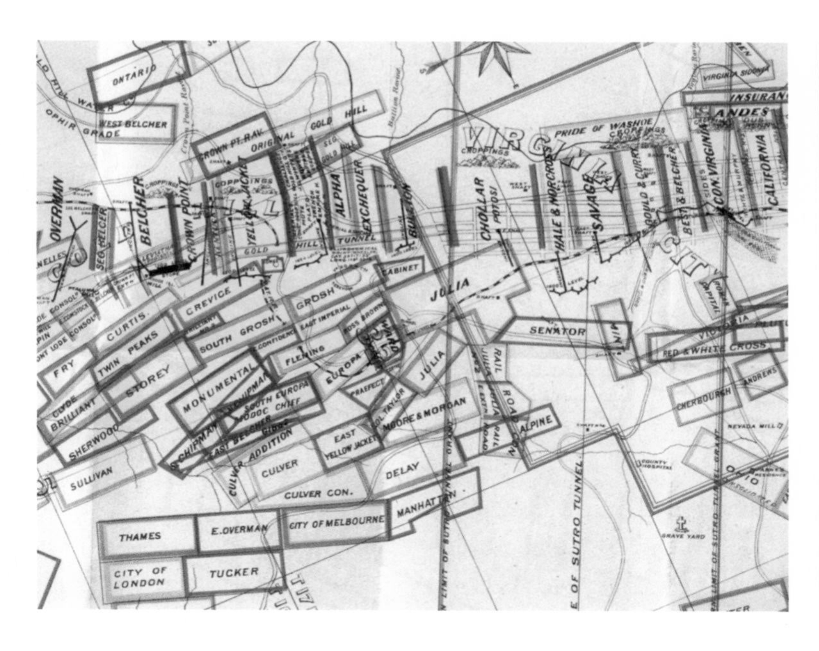

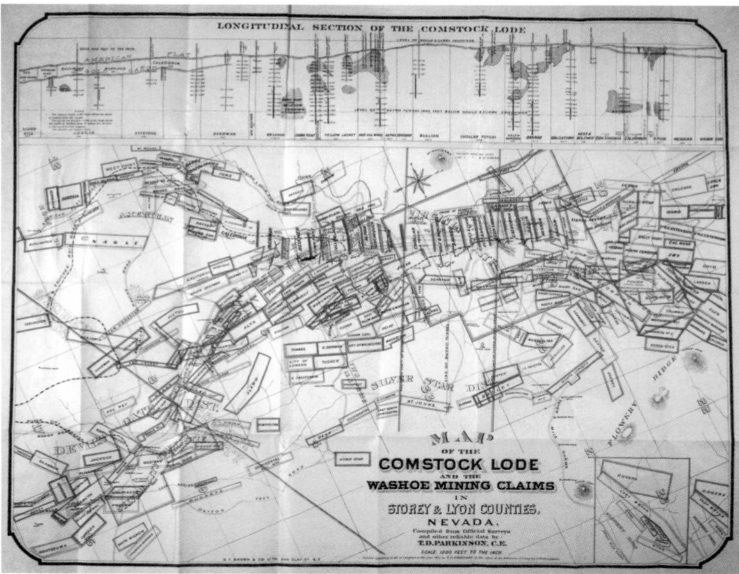

Reaching for Gold in the Black Hills

CREATORS: Chicago and North Western Railway (publisher), Rand, McNally & Co. (mapmaker and printer)

TITLE: *Gold Fields of the Black Hills of Dakota* (Chicago, Rand, McNally & Co., printers [c. 1877])

SIZE: 24 x 7.1 inches

LOCATION: Newberry Graff 690

CREDIT: Gift of Everett D. Graff

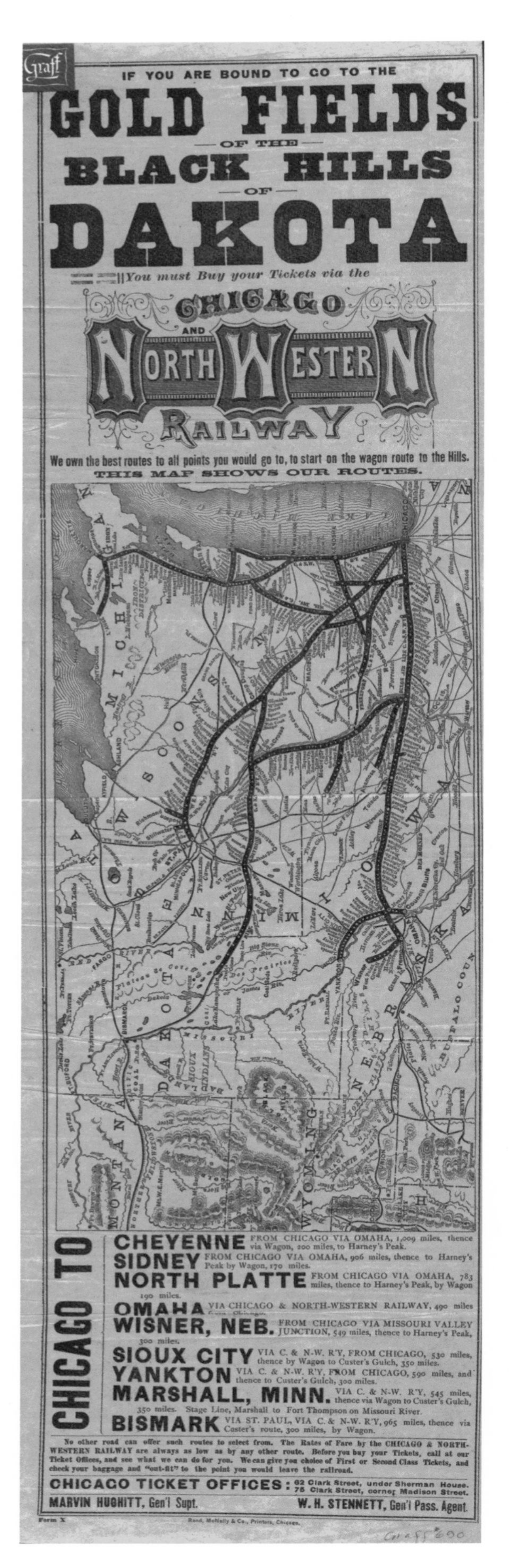

The 1874 report of gold in the Black Hills transformed this corner of the Dakota Territory into a hotly contested space. A holy site for peoples native to the northern Great Plains, the land was set aside as reservation by the U. S. government in 1867. The Indians knew about the gold but did not exploit it. When Lieutenant Colonel George A. Custer's (1839–1876) Army expedition discovered the gold, and the *Chicago Inter-Ocean* broadcasted its presence (on August 24, 1874), white prospectors flocked to the region. The government soon pressed the Indians to cede or sell the land back. When they refused, the government ordered that they halt their annual wanderings and confine themselves to the reservation, sending in soldiers to enforce the new regulation. Outraged by this invasion of their territory, the Indians responded violently. On June 25, 1876, Cheyenne and Lakota warriors defeated troops commanded by Custer in the Battle of Little Big Horn. The Indians won the battle but lost the war, surrendering to U.S. troops in the fall of 1877.

Eager to sell train tickets to miners, Western railroads ignored the conflict; their maps also downplayed the remoteness of the Black Hills and the presence of rival lines. As the area was not served by any of the main transcontinental routes, several lines competed to carry passengers to the ends of their spurs in the northern Great Plains. From these points, passengers would have to make the rest of the journey by wagon.

The Chicago and North Western Railway (C&NW) made their line appear more convenient by deftly manipulating the region's geography. The mapmaker contracted the width of Dakota, making it appear about as wide as Minnesota and Wisconsin. As a result, the C&NW lines from Lake Michigan seem to run much closer to the Black Hills than they actually did. The map includes the Northern Pacific and Union Pacific lines, which the C&NW relied on for western connections to the area. But the cartographer highlighted the C&NW routes by drawing them more boldly and locating them at the center of the map. The C&NW incorporated the same map into several publications. In addition to the broadside shown here, it appeared in a rate bulletin, combined with timetables and ticket prices, published on April 10, 1878.

References: Akerman 2000; Akerman 2003; Chicago and North Western Railway 1878; Greever 1963; Peters 1984; White 1991a

Stimulating Tourism

The earliest promoters of tourism in the West saw its spectacular landscape as a natural resource to be exploited, much as farmers exploited its soil and miners its minerals. Like agriculture and mining, tourism developed through a combination of public and private initiatives. Whereas most scenic destinations in the East were privately owned, the recreational lands in the West were chiefly in the public domain. Because the most dramatic landscapes were often remote, they remained largely unknown and inaccessible until government surveys produced maps, descriptive reports, and popular articles. The railroads followed on the heels of the explorers, extending their lines, building hotels, lobbying for the establishment of national parks, and advertising scenic places to attract passengers. Although scientific study of the topography and geology of the wilderness may seem at odds with the development of tourism, early explorers recognized that promoting natural wonders to politicians and leisure travelers could help insure the landscape's preservation.

4.12

Mapping Nature's Wonderland

CREATORS: Carl J. Hals and Arvid Rydström

TITLE: *Map of the Yellowstone National Park: Compiled from Different Official Explorations and Our Personal Survey*, 1882; *Alice's Adventures in the New Wonderland* (Chicago: Poole Bros., [1884])

SIZE: 22 x 18 inches

LOCATION: Newberry Map4F G4262 Y4 1884 H3

The U. S. Congress set aside the northwest corner of what is now Wyoming as Yellowstone National Park in 1872. The prospect of mineral wealth had attracted the attention of officials of the Montana Territory and the Northern Pacific Railroad to the area in the late 1860s. The region caught the eye of a broader public after Charles W. Cook (1839–1927), David E. Folsom (1839–1918), and William Peterson (1834–1919) explored the territory in 1869. Their descriptions stressing the "awful grandeur and sublimity" of the landscape appeared the following summer in a Chicago magazine, *The Western Monthly*, after *The New York Times*, *Scribner's*, and *Harper's* dismissed the account as unbelievable. In the summer of 1871, Ferdinand Vandeveer Hayden (1829–1887) directed his government survey of the Western territories to Yellowstone, sending in several scientists, a topographer, a photographer, and two artists. Hayden quickly recognized the region's singular charms, and that fall he energetically supported the campaign to establish it as the first national park. Officials and investors connected to the Northern Pacific Railroad supported his efforts, envisioning a tourist resort served exclusively by their line. By 1883 the railroad had extended its tracks to Cinnabar, near the park's northern entrance in Montana. The railroad also financed the construction of the National Hotel (later renamed Mammoth Hot Springs Hotel), the park's first, which opened in 1884 at Mammoth Hot Springs.

The Northern Pacific issued *Alice's Adventures in the New Wonderland* in 1884 to launch its first full season of tourist service to Yellowstone. Hals and Rydström (1857–after 1919), civil engineers employed by the railroad, compiled the map that fills one side of the brochure, relying largely on the map Henry Gannett (1846–1914) published in 1882 based on Hayden's 1878 expedition into the park. Hals and Rydström included more topographical detail than most promotional railroad maps, translating Gannett's contour lines into hachures to make the relief readily understood. Such cartographic precision enhanced the map's promotional value, assuring audiences that Yellowstone's sublime wonders were well-documented realities rather than topographic fantasies.

The brochure leavens the map's scientific tone with the imaginative text printed on the opposite side. The cover panel pictures the fictional protagonist, named after the heroine of Lewis Carroll's *Alice in Wonderland*. Written in the form of a letter to imitate an authentic, personal account, the text stresses the comforts of the train and hotel as much as the glory of the geysers, hot springs, and other "natural curiosities." An articulate and sophisticated English traveler, Alice lends feminized civility and domesticity to a landscape known as "bubbling hell," dominated by men as rugged as the topography. Yellowstone remained the centerpiece of the Northern Pacific's tourist literature for decades, becoming the focus of its annual booklet, *Wonderland*, in the late 1880s. The line commanded the Yellowstone tourist trade into the early twentieth century, when competing railroads arrived from both the east and west.

References: Akerman 2000; Akerman 2003; Cook 1870; Fifer 1988; Haines 1977; Hyde 1990; Runte 1994

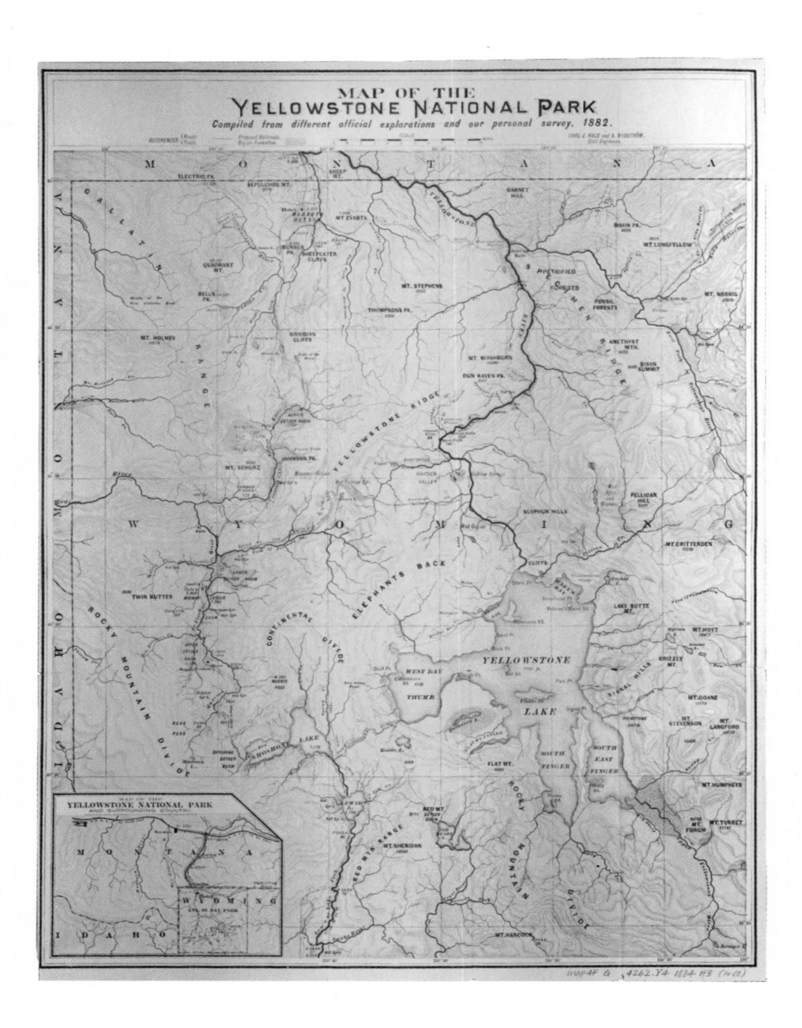

4.13

The Switzerland of America

CREATORS: James Steele (brochure author), Coella Lindsay Ricketts (map draftsman)

TITLE: *Colorado* (Chicago: Passenger Department, Chicago, Burlington & Quincy Railroad Co., 1900)

SIZE: 8 x 6.25 inches (booklet, closed); 8 x 12 inches (booklet, two-page spread); 19.25 x 11 inches (map)

LOCATION: Newberry CB&Q MS, box 1, folders 19:15, 19:16, 19:17

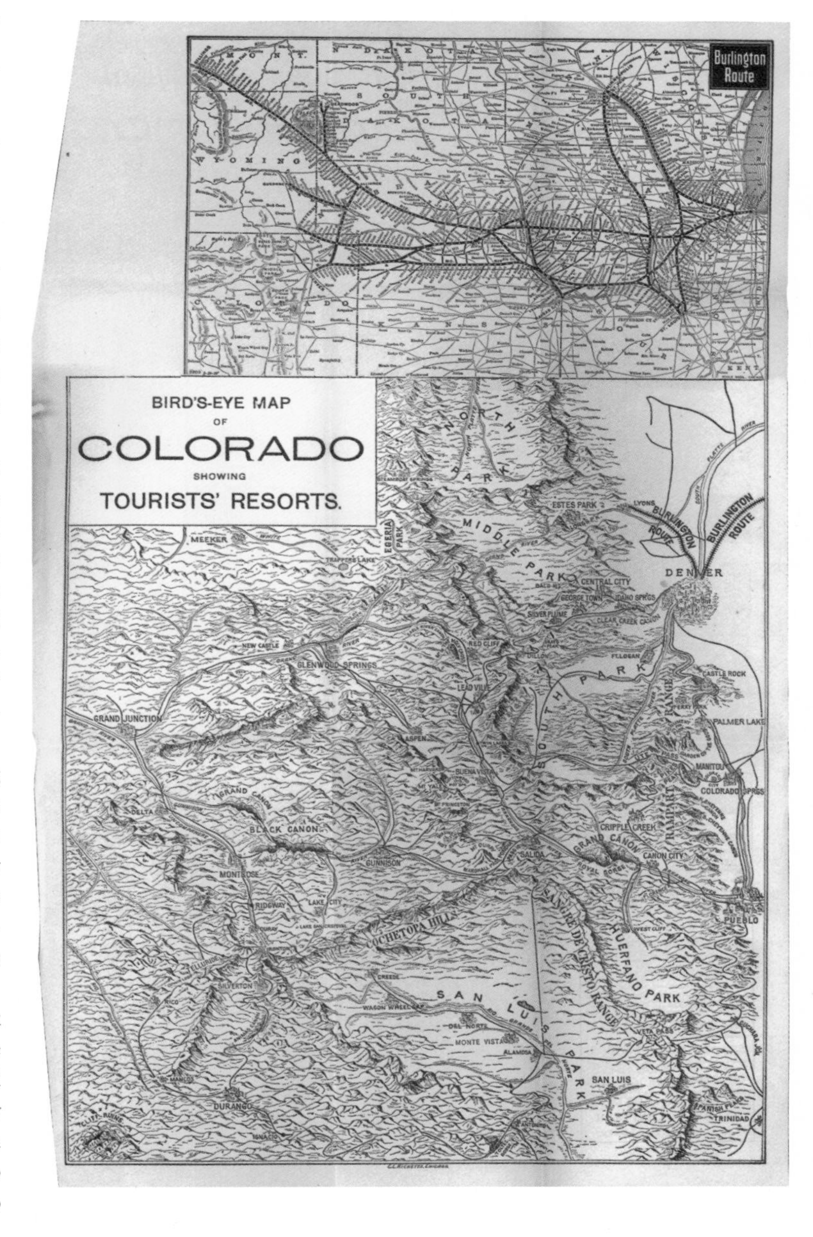

The Chicago, Burlington & Quincy Railroad (CB&Q) followed the Northern Pacific's lead in promoting the West's majestic scenery to tourists. In a booklet issued in several editions between 1900 and 1905, copywriter James Steele also described Colorado's pleasure grounds as a "Wonderland," boasting that the mountains of central Colorado were grander than the Swiss Alps.

The cover illustration summarized Colorado's tourist draws: the Indian panning for gold alluded to its colorful past, enriched by mining lore and Native American traditions, while the snow-capped peak attested to its timeless natural beauty. The scenery Steele extolled lacked Yellowstone's novelty: mining rushes had drawn publicity, emigrants, and railroads to the Colorado mountains by the mid-nineteenth century. After the mines were tapped out, the railroads aimed to fill their trains with sightseers, recasting mining towns as tourist attractions. Steele noted that the mining village of Idaho Springs was "also a health resort," while Silver Plume now served as the destination of leisure travelers on the high-altitude "Loop" railway.

Inside the booklet, photographs served as the primary means of captivating potential visitors. The images are credited to the Detroit Photographic Company, one of the largest publishers of photographic views in the early twentieth century. The pictures may have been taken by the celebrated photographer William Henry Jackson (1843–1942) who added his negatives of scenes along railroad lines to the company's inventory in 1897. The photographs intersperse views of breathtaking scenery with engineering wonders built by the railroad, such as bridges over deep canyons, tunnels through mountains, and steeply inclined tracks.

Because the photographs took pride of place, the map folded into the back of *Colorado* was more rudimentary than the one the NPRR used for its Alice brochure (see Item 4.13). Cheaply printed on thin paper, the sheet juxtaposed a standard map of the Burlington's routes between Chicago and the Rocky Mountains with a map of Colorado showing the mountains and resorts from a bird's eye perspective. This crude map was drawn by a leading Chicago calligrapher, Coella Lindsay Ricketts (1859–1941), who probably took the job when he was between more artistic assignments.

References: Akerman 2003; Musich 2006

C O L O R A D O

Chapter Four

The Colorado Pleasure Grounds, II.
The "Loop" Journey.

IN ALMOST the opposite direction from that which takes the visitor to Manitou, and still near the eastern rim of the mountains, is one cañon that can be visited from Denver in a day. It lies upon one of the lines of the U. P., D. & G. road, and is known as Clear Creek Cañon.

This defile in the mountains has been long known. When Colorado was young it was a miners' wagon-road over the range. Tens of thousands have seen it, and it still remains, especially to one who has no time to see the overpowering scenery in the interior, an experience not to be left out.

From Denver it is fifteen miles to Golden; which is the simple name of a town, one of the original gold camps; over a stretch of country that was once an inlet of the great plains sea, and is as level as the Nebraska prairie. It is a fruit, farm and ranch country, suggesting nothing of the scene that is so near at hand. The basin in which the town lies is the bottom of this sea, and the rocks around the shore are water-worn, and show where the waves once lapped. But a little distance beyond lies the opening to this famous gorge and its tumbling stream, and thence the road follows the cañon for more than twenty miles.

The place, like most cañons, is apparently a cooling-crack; a place that opened in the shrinking of the crust when the white-hot world began to harden. The projections of one side vaguely fit the indentations of the other. Very often the red walls come very close together, revealing, as one looks upward, only a narrow blue streak where the sky is.

39

Chicago Corners Commercial Mapping

The cartographic industry in Chicago matured in tandem with the city itself, mirroring its breathtaking rise in six short decades from a frontier outpost to the nation's second-largest metropolis. By the end of the nineteenth century, Chicago had eclipsed Philadelphia and New York to become the national center of map publishing. Chicago cartography flourished by exploiting the commercial potential of maps. Mapmakers took advantage of the city's position as the primary supplier of goods to the entire West, providing individual consumers and businesses throughout the region with inexpensive, utilitarian maps. By the last quarter of the nineteenth century, Rand McNally & Co.'s innovative manufacturing, product development, and marketing enabled the firm to dominate the map industry not only in the West, but in much of the nation.

4.14

The Cartography of Insurance

CREATORS: Frederic Cook (surveyor, draftsman), B. W. Phillips & Co. (publishers)

TITLE: *Fire Insurance Maps of Chicago, Illinois* (Chicago: B. W. Phillips & Co., 1865)

SIZE: 16.5 x 13.6 inches

LOCATION: Chicago History Museum, Chicago, IL

Fire insurance underwriters developed specialized maps to enable them to assess the vulnerability of urban and suburban buildings without inspecting the sites first hand. The Phoenix Assurance Company introduced fire insurance maps in London in the late eighteenth century, but the genre flourished most extensively in the United States and Canada. In the second half of the nineteenth century, the Sanborn Map Company gradually monopolized the trade in the U.S. Before Sanborn standardized the formula, local cartographers supplied underwriters with varied fire insurance maps and atlases.

Frederic Cook's 1865 compilation of maps provides a detailed picture of Chicago's built environment before the Great Fire of 1871. Although later atlases included residential neighborhoods, Cook focused exclusively on commercial and industrial structures. The volume's first section pictures Chicago's downtown business district, while the second focuses on industrial facilities, featuring full-page diagrams of grain elevators, slaughtering and meat packing houses, and breweries. These maps document the industries that propelled the city's meteoric growth. Farmers and ranchers throughout the West sent grain, hogs, and cattle to Chicago, where they were processed into commodities and shipped to Eastern consumers.

The sheet showing the block between Wabash and State, Washington and Randolph Streets illustrates the customary color-coding used for insurance maps, indicating brick buildings in pink, wood in yellow, stone in gray. Also typical of insurance maps is the updating by hand. The page includes several annotations in pencil, and identifies in ink the lot that real estate mogul Potter Palmer developed for the department store he founded, then known as Field, Leiter & Co. Cook's maps are unusual for their inclusion of architectural details such as the Greek pediments and cornice dentils defining the roof of the Garrit block.

References: Library of Congress 1981; Ristow 1968; Rowley 1984

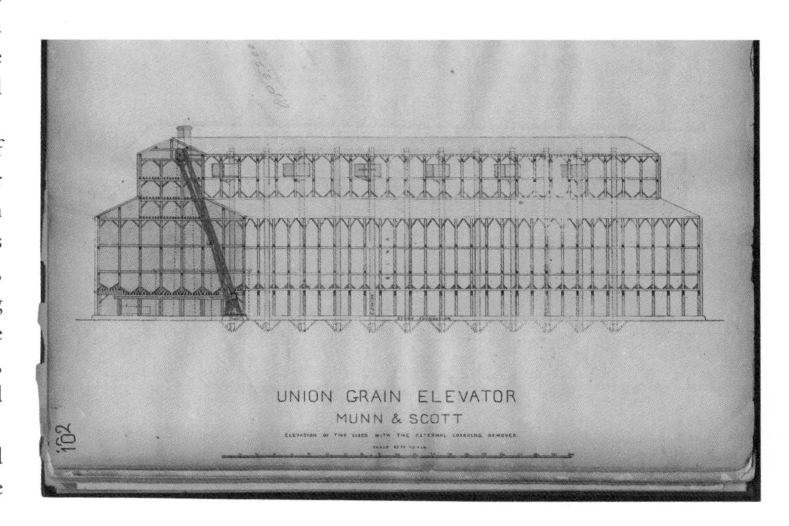

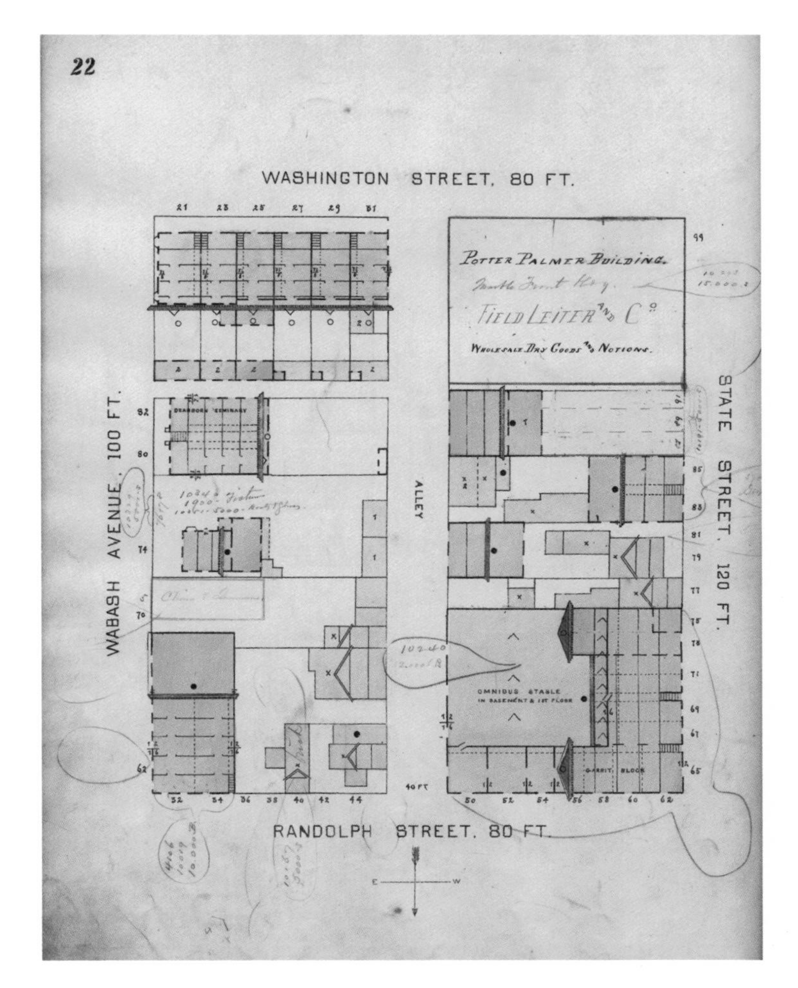

4.15, 4.16, 4.17, 4.18

Scaled to Sell: Maps for Every Need

[4.15] CREATOR: Rand McNally and Company

TITLE: *Idaho*, in *Business Atlas, Containing Large Scale Maps of Each State and Territory of the Great Mississippi Valley and Pacific Slope* (Chicago: Rand McNally & Co., 1876–1877)

SIZE: 20.5 x 13.75 inches

LOCATION: Newberry Map 4F G3700 1876 R3

CREDIT: Gift of Rand McNally & Co.

[4.16] CREATOR: Rand McNally and Company

TITLE: *Business Atlas, Containing Large Scale Maps of Each State and Territory of the United States, the Provinces of Canada, West India Islands, Etc., Etc.* (Chicago: Rand McNally & Co., 1880)

SIZE: 14.75 x 12 inches

LOCATION: Newberry oversize RMcN Atlas C7 1880

CREDIT: Gift of Rand McNally & Co.

William H. Rand
CREDIT: Rand McNally and Company Records, Midwest Manuscript Collection, Newberry Library, Chicago.

Andrew McNally
CREDIT: Rand McNally and Company Records, Midwest Manuscript Collection, Newberry Library, Chicago.

By the turn of the twentieth century, the name Rand McNally was nearly synonymous with American mapmaking. The firm's founders, William H. Rand (1828–1915) and Andrew McNally (1836–1904), both arrived in Chicago in the 1850s. Rand hired McNally to work in his printing office, which flourished by serving the city's signature enterprises, especially the railroads. The firm specialized in railroad tickets, annual reports, and stationery before introducing the *Western Railway Guide* in 1871 and then, eighteen months later, moving into map publishing.

Rand McNally imitated the industrial methods of the corporations it served. The company exploited the potential of cerography (wax engraving), a process that utilized an assembly line and allowed printers to combine type and graphics on the same sheet. Because wax-engraved plates were more durable than copperplates, the publisher could print maps in great quantities at low costs. Rand McNally compensated for wax-engraving's cruder lines with clean design, rejecting the busy layouts and fussy typefaces favored by competitors. In creating such a distinctive look for their maps, Rand McNally furthered consumers' ready identification of the brand.

The firm also utilized up-to-date marketing, adapting their standardized maps to a wide range of products for different audiences. The *Business Atlas*, first issued in 1876, catered to salesmen and managers who wanted to know the sizes and

[4.16]

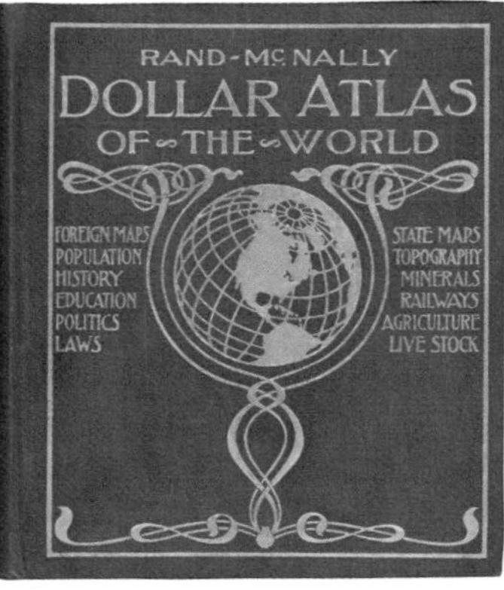

[4.18]

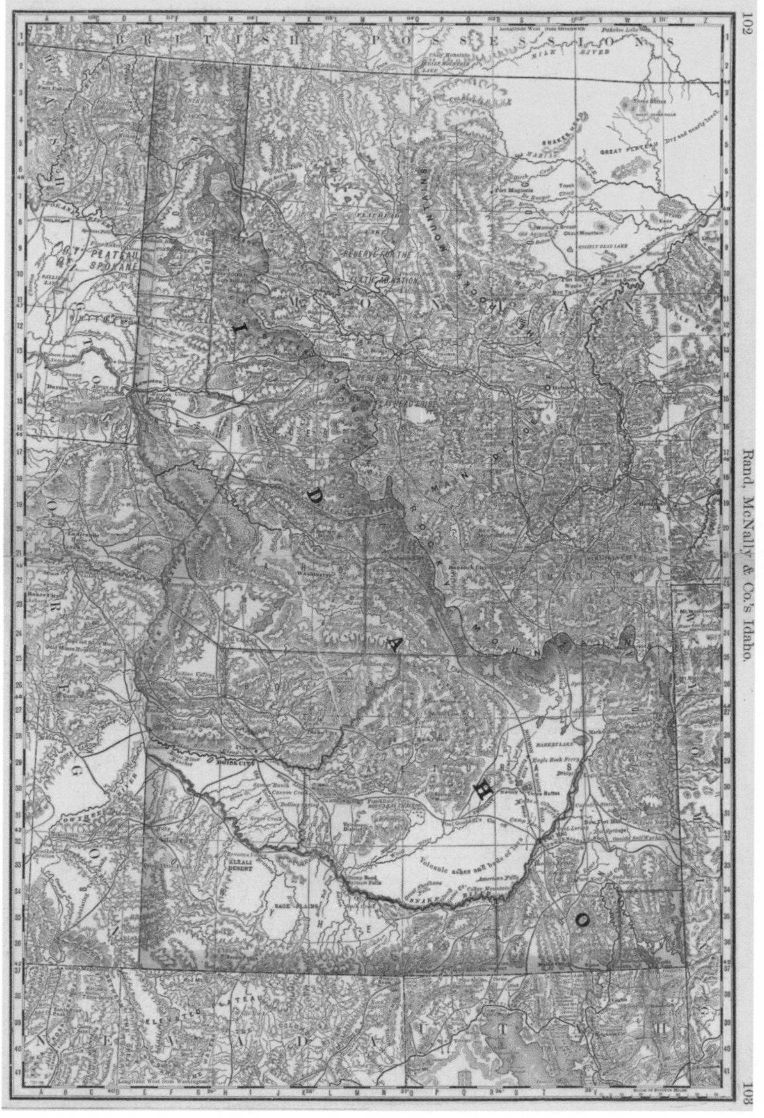

[4.15]

[4.17] CREATOR: Rand McNally and Company

TITLE: *Rand McNally & Company's Standard Map of the United States of America, With Portions of the Dominion of Canada and the Republic of Mexico* (Chicago: Rand McNally & Co., 1887)

SIZE: 107.2 x 179.2 inches

LOCATION: Newberry Ayer 133 R18 1887

CREDIT: Gift of Edward E. Ayer

[4.18] CREATOR: Rand McNally and Company

TITLE: *Rand McNally Dollar Atlas of the World* (Chicago: Rand McNally & Co., 1918)

SIZE: 7.25 x 6.25 inches

LOCATION: Newberry R McN Atlas 1918 copy 2

CREDIT: Gift of Rand McNally & Co.

locations of towns and their proximity to transportation lines, but did not need topographical or scientific information. In contrast to deceptive railroad maps (see Items 4.5 and 4.11), the *Business Atlas* succeeded because it was accurate; it continues today as the *Commercial Atlas and Marketing Guide*.

In 1900 Rand McNally met consumer demand for an inexpensive, handy reference book with the *Dollar Atlas of the World*. The volume combined features of earlier geography textbooks and family atlases (see Items 3.1 through 3.6). In 1887 the firm issued a giant (nine by fourteen feet) map of the United States to satisfy a more extravagant clientele and to display in public places such as offices, stores, schools, and railroad stations. Brandishing Rand McNally's name in a prominent cartouche, the map advertised the patriotism and big ambitions of its makers as well as its owners. Although the content of these Rand McNally products was not specific to the West, they embodied the entrepreneurial spirit and utilitarian style that helped define the region in the popular imagination.

References: Dillon 2003; Dillon 2007; Peters 1984; Schulten 2001

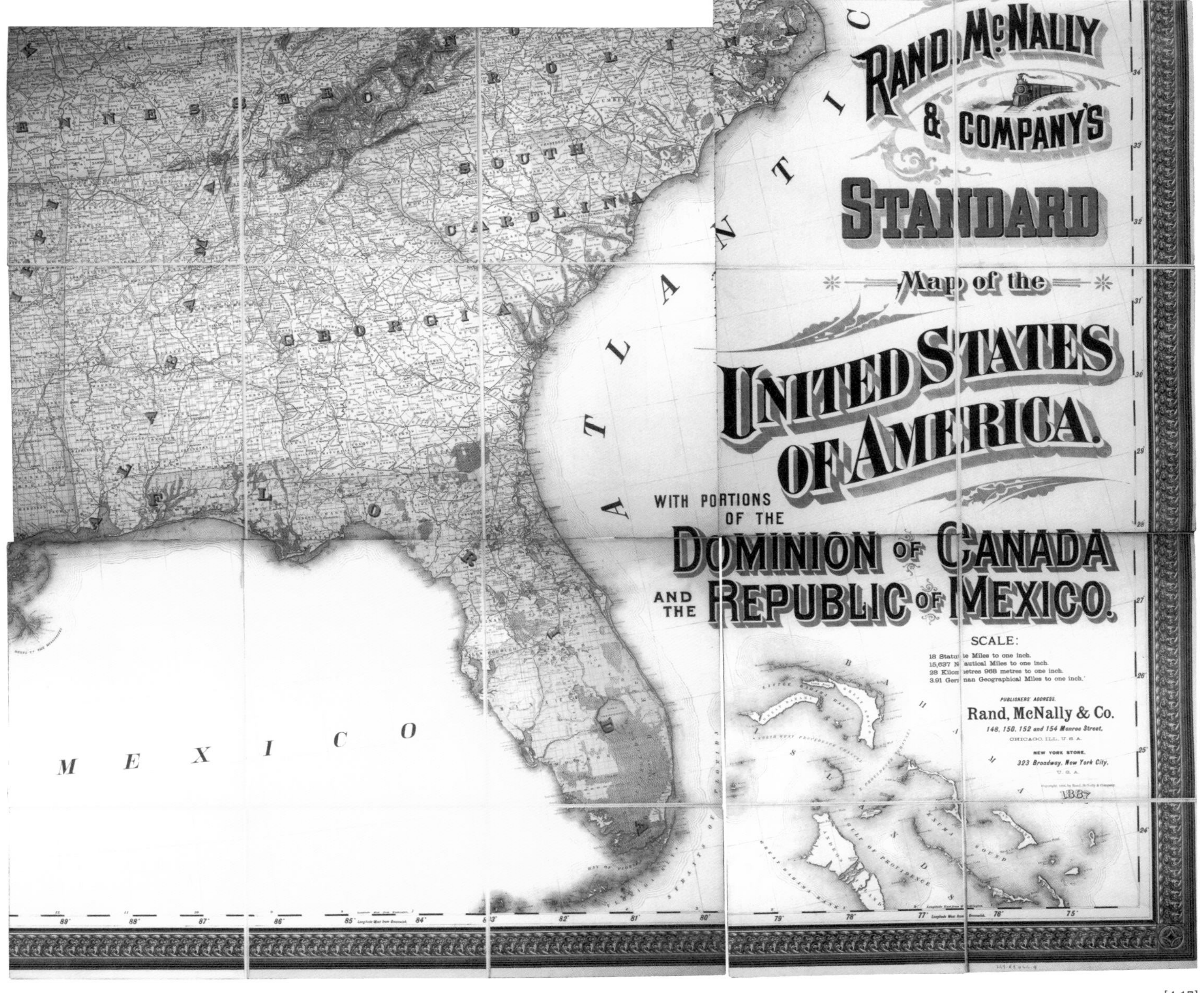

[4.17]

Bibliography

Akerman, James R. 2000. "Riders Wanted: Maps as Promotional Tools in the American Transportation Industry." Paper presented at *Maps and Popular Culture*, the Second Biennial Virginia Garrett Lectures on the History of Cartography, Arlington, Texas, October.

———. 2003. "Mapping a Nation on Rails: Railroad Cartography and American Identity, 1865–1941." Paper presented at the 20th International Conference on the History of Cartography, Cambridge, MA and Portland, ME, June.

———, ed. 2006. *Cartographies of Travel and Navigation*. Chicago: University of Chicago Press.

———. 2007. "Mapping 'Wonderland:' Explorers, Tourists and the Cartography of Yellowstone National Park." 22nd International Conference on the History of Cartography, Bern, Switzerland, July.

Akerman, James. R. and Robert W. Karrow, Jr., eds. 2007. *Maps: Finding Our Place in the World*. Chicago and London: University of Chicago Press.

Alekseev, A. I. 1987. *The Odyssey of a Russian Scientist: I. G. Voznesenskii in Alaska, California and Siberia, 1839–1849*. Trans. Wilma C. Follette, ed. R. A. Pierce. Kingston, Ontario: The Limestone Press.

———. 1990. *The Destiny of Russian America, 1741–1867*. Trans. Marina Ramsey, ed. R. A. Pierce. Kingston, Ontario: The Limestone Press.

Allen, John Logan. 1975. *Passage Through the Garden: Lewis and Clark and the Image of the American Northwest*. Urbana: University of Illinois.

Ambrose, Stephen E. 1996. *Undaunted Courage: Meriwether Lewis, Thomas Jefferson, and the Opening of the American West*. New York: Simon & Schuster.

———. 2000. *Nothing Like It in the World: The Men Who Built the Transcontinental Railroad, 1865–1869*. New York: Simon & Schuster.

Anderson, Fred. 2000. *The Crucible of War: The Seven Years' War and the Fate of Empire in British North America, 1754–1766*. New York: Alfred A. Knopf.

———. 2005. *The War That Made America: A Short History of the French and Indian War*. New York: Viking.

Andrews, C. L. 1922. *The Story of Sitka*. Seattle: Lowman & Hanford Co.

———. 1938. *The Story of Alaska*. Caldwell, Idaho: Caxton Printers, Ltd.

Angle, Paul. 1952. "The First Published Map of Chicago." *Chicago History* vol. 3 (Spring): 89–93.

Anon. 1760. "Successful Expedition of Governor Lyttleton against the Cherokee Indians, with an Accurate Map of their Country." *London Magazine* vol. 29 (February): 95–96.

Arndt, Katherine L., and Richard A. Pierce, 2003. *Sitka National Historical Park Historical Context Study: A Construction History of Sitka, Alaska, as Documented in the Records of the Russian-American Company*, 2nd ed. [Sitka, Alaska]: Sitka National Historical Park, National Park Service.

Bancroft, Caroline. 1958. *Gulch of Gold: A History of Central City, Colorado*. Boulder: Johnson Publishing Company.

Bancroft, Hubert Howe. 1890. *History of Nevada, Colorado, and Wyoming, 1540–1888*. San Francisco: The History Company.

Barnard, Henry. 1859. *Educational Biography, Memoirs of Teachers, Educators*. New York: F. C. Brownell.

Bartlett, Richard A. 1962. *Great Surveys of the American West*. Norman: University of Oklahoma Press.

Belyea, Barbara. 1992. "Amerindian Maps: The Explorer as Translator." *Journal of Historical Geography* 18 (July): 267–77.

Berkeley, Edmund, and Dorothy Smith Berkeley. 1974. *Dr. John Mitchell: The Man Who Made the Map of North America*. Chapel Hill: University of North Carolina Press.

Berton, Pierre. 1970. *The National Dream*. Toronto: McClelland & Stewart.

Black, Lydia T. 2004. *Russians in Alaska, 1732–1867*. Fairbanks: University of Alaska Fairbanks.

Bollaert, William. 1956. *William Bollaert's Texas*, ed. W. Eugene Hollon and Ruth Lapham Butler. Norman: University of Oklahoma Press, 1956.

Bosse, David. 1986. "The Maps of Robert Rogers and Jonathan Carver." *American Magazine and Historical Chronicle* 2 (Spring/Summer): 45–61.

Bourne, Edward Gaylord. 1906. "The Travels of Jonathan Carver." *The American Historical Review* 11, no. 2 (January): 287–302.

Brown, Dee. 1977. *Hear That Lonesome Whistle Blow: Railroads in the West*. New York: Holt, Rinehart and Winston.

Bryant, Keith L, Jr. 1974. *History of the Atchison, Topeka and Santa Fe Railway*. New York: Macmillan.

Buck, Solon Justus. 1914. *Travel and Description, 1765–1865; Together with a List of County Histories, Atlases, and Biographical Collections and a List of Territorial and State Laws*. Illinois Historical Collections, IX. Springfield, IL: The Trustees of the Illinois State Historical Library.

Buckley, Eleanor Claire. 1911. "The Aguayo Expedition into Texas and Louisiana, 1719–1722." *Quarterly of Texas State Historical Association* 15 (July): 1–65.

Buisseret, David. 1991. *Mapping the French Empire in North America. An Interpretive Guide to the Exhibition Mounted at the Newberry Library on the Occasion of the Seventeenth Annual Conference of the French Colonial Historical Society (La Société d'Histoire Coloniale Française)*. Chicago: The Newberry Library.

———, ed. 2007. *The Oxford Companion to World Exploration*, 2 vols. Oxford: Oxford University Press.

Burrows, J. M. D. 1888. *Fifty Years in Iowa: Personal Reminiscences of J. M. D. Burrows*. Davenport, Iowa: Glass and Company.

Campbell, Edward Gross. 1938. *The Reorganization of the American Railroad System, 1893–1900*. New York: Columbia University Press.

Campbell, Tony. 1987. "Portolan Charts from the Late Thirteenth Century to 1500." In *The History of Cartography*, ed. J. B. Harley and David Woodward. Chicago and London: University of Chicago Press, vol. 1, 371-463.

Carver, Jonathan. 1778. *Travels Through the Interior Parts of North-America in the Years 1766, 1767, and 1768*. London: printed for the author.

Chicago and North Western Railway. 1878. *To the Black Hills, Rate Bulletin no. 15...* Chicago: Chicago and North Western Railway Company.

Churchill, Charles B. 1991. "Thomas Jefferson Farnham: An Exponent of American Empire in Mexican California." *The Pacific Historical Review* 60, no. 4 (November): 517–537.

Cloud, John. 2007. "The 200th Anniversary of the Survey of the Coast." *Prologue* (Spring): 24–33.

Cohen, Paul E. 2002. *Mapping the West: America's Westward Movement, 1524–1890*. New York: Rizzoli.

Conzen, Michael P., ed. 1984a. *Chicago Mapmakers: Studies in the Rise of the City's Map Trade*. Chicago: Chicago Historical Society.

———. 1984b. "The County Landownership Map in America. Its Commercial Development and Social Transformation, 1814–1939." *Imago Mundi* 36:9–31.

———. 1984c. "Maps for the Masses: Alfred T. Andreas and the Midwestern County Atlas Trade." *In Chicago Mapmakers: Studies in the Rise of the City's Map Trade*, ed. Michael P. Conzen, 46–63. Chicago: Chicago Historical Society for the Chicago Map Society.

———. 1997. "The All-American County Atlas: Styles of Commercial Landownership Mapping and American Culture." In *Images of the World: The Atlas Through History*, ed. John A. Wolter and Ronald E. Grim, 331–365. Washington, DC: Library of Congress.

Cook, Charles W. 1870. "The Valley of the Upper Yellowstone." *Western Monthly Magazine* 4 (July): 60–67.

Cronon, William. 1991. *Nature's Metropolis: Chicago and the Great West*. New York: W. W. Norton.

Cronon, William, et al., eds. 1992. *Under an Open Sky: Rethinking America's Western Past*. New York and London: W. W. Norton & Company.

Cumming, William P. 1974. *British Maps of Colonial America*. Chicago and London: University of Chicago Press.

———. 1998. *The Southeast in Early Maps*. 3rd ed. / rev. and enl. by Louis De Vorsey, Jr. Chapel Hill: University of North Carolina Press.

Cumming, William P., R. A. Skelton, and D. B. Quinn. 1972. *The Discovery of America*. New York: American Heritage Press.

Cumming, William P., et al. 1974. *The Exploration of North America, 1630–1776*. New York: G. P. Putnam's Sons.

Dale, Edward Everett. 1949. *Oklahoma, the Story of a State*. Evanston, IL: Row, Peterson.

Danzer, Gerald. 1984a. "Chicago's First Maps." In *Chicago Mapmakers: Studies in the Rise of the City's Map Trade*, ed.

Michael P. Conzen, 12–22. Chicago: Chicago Historical Society for the Chicago Map Society.

———. 1984b. "George F. Cram and the American Perception of Space." In *Chicago Mapmakers: Studies in the Rise of the City's Map Trade*, ed. Michael P. Conzen, 32–45. Chicago: Chicago Historical Society for the Chicago Map Society.

Darnell, Regna Diebold. 1969. "The Development of American Anthropology 1879–1920: From the Bureau of American Ethnology to Franz Boas." Ph.D. Diss., University of Pennsylvania.

———. 1971. "The Powell Classification of American Indian Languages." *Papers in Linguistics* 4, no. 1 (July), pp. 71–110.

Dawdy, Shannon Lee. 2008. *Building the Devil's Empire: French Colonial New Orleans*. Chicago: University of Chicago, 2008.

Deverell, William F. 1994. *Railroad Crossing: Californians and the Railroad, 1850–1910*. Berkeley: University of California Press.

Dillon, Diane. 2003. "Mapping Enterprise: Cartography and Commodification at the 1893 World's Columbian Exposition," in *Nineteenth-Century Geographies: Anglo-American Tactics of Space*, ed. Helena Michie and Ronald Thomas. New Brunswick: Rutgers University Press.

———. 2007. "Consuming Maps," in *Maps: Finding Our Place in the World*, ed. James R. Akerman and Robert W. Karrow, Jr. Chicago and London: University of Chicago Press.

Dumont de Montigny, Jean-François-Benjamin. 2008. *Les Mémoires de Dumont de Montigny. Aventures au monde atlantique, 1715–1747*, ed. Carla Zecher, Gordon M. Sayre, and Shannon Lee Dawdy. Québec: Les Éditions du Septentrion, 2008.

Dunnigan, Brian Leigh. 2001. *Frontier Metropolis: Picturing Early Detroit, 1701–1838*. Detroit: Wayne State University Press.

Durrie, Daniel S. 1872. "Captain Jonathan Carver and 'Carver's Grant.'" *Wisconsin Historical Collections* 6 (1872): 220–270.

Edney, Matthew H. 1986. "Politics, Science, and Government Mapping Policy in the United States, 1800–1925." *The American Cartographer* 13 (1986): 295–306.

———. 1997. "Introduction and Overview," in *John Mitchell's Map: An Irony of Empire*. Web site, Osher Map Library and Smith Center for Cartographic Education, University of Southern Maine. http://www.usm.maine.edu/~maps/mitchell/intro.html.

———. 2007. "A Publishing History of John Mitchell's Map of North America, 1755–1775." *Cartographic Perspectives* no. 58 (Fall). Forthcoming.

———. 2008. "John Mitchell's Map of North America (1755) in its Official and Public Contexts." *Imago Mundi* 60, no. 1. Forthcoming.

Ellis, Joseph. 2004. *His Excellency, George Washington*. New York: Alfred A. Knopf.

Falk, Marvin W. 1990. "Mapping Russian America, 1760–1867." In *Russia in North America: Proceedings of the Second International Conference on Russian America*. Kingston, Ontario: The Limestone Press.

Farnham, Thomas Jefferson. 1849. *Life, Adventures, and Travels in California*. New York: Nafis & Cornish; St. Louis, MO: Van Dien & Macdonald.

Farnham, Wallace D. 1965. "Grenville Dodge and the Union Pacific: A Study of Historical Legends." *Journal of American History* 51 (March): 632–650.

Fifer, J. Valerie. 1988. *American Progress: The Growth of the Transport, Tourist, and Information Industries in the Nineteenth-Century West, Seen Through the Life and Times of George A. Crofutt, Pioneer and Publicist of the Transcontinental Age*. Chester, CT: Globe Pequot Press.

Fogelson, Raymond D., ed. 2004. *Handbook of North American Indians*, Volume 14: Southeast, general ed. William Sturtevant. Washington, DC: Smithsonian Institution.

Frémont, John Charles. 1843. *Report on an Exploration of the Country Lying between the Missouri River and the Rocky Mountains on the Line of the Kansas and Great Platte Rivers*. Washington: Printed by Order of the United States Senate.

———. 1845. *Report of the Exploring Expedition to the Rocky Mountains in the Year 1842, and to Oregon and North California in the Years 1843–44*. Washington: Gales and Seaton.

Gibson, James. 1976. *Imperial Russia in Frontier America: The Changing Geography of Supply of Russian America, 1784–1867*. New York: Oxford University Press.

Gilpin, William. 1860. *The Central Gold Region. The Grain, Pastoral and Gold Regions of North America. With Some New Views of its Physical Geography; and Observations on the Pacific Railroad*. Philadelphia: Sower, Barnes & Co.; St. Louis: E. K. Woodward.

Glover, Lyman B. 1911. *Charles Edward Kohl: A Biography and Appreciations*. [Chicago?]: Privately printed.

Goetzmann, William H. 1959. *Army Exploration in the American West, 1803–1863*. New Haven: Yale University Press.

———. 1966. *Exploration and Empire: The Explorer and the Scientist in the Winning of the American West*. New York: Alfred A. Knopf.

———. 1979. "Limner of Grandeur: William Henry Holmes." *American West* 15 (May–June): 20–21, 61–63.

———. 1986. *The West of the Imagination*. New York: W. W. Norton and Co.

Golder, F. A. 1914. *Russian Expansion of the Pacific, 1641–1850*. New York: Paragon Books, 1971.

Gordon, Sarah H. 1996. *Passage to Union: How the Railroads Transformed American Life, 1829–1929*. Chicago: Ivan R. Dee, 1996.

Greever, William S. 1963. *The Bonanza West: The Story of the Western Mining Rushes, 1848–1900*. Norman: University of Oklahoma Press.

Gregory, John Goadby. 1896. *Jonathan Carver: His Travels in the Northwest in 1766-8*. Milwaukee, WI: Parkman Club Publications no. 5.

Grodinsky, Julius. 1962. *Transcontinental Railroad Strategy, 1869–1893: A Study of Businessmen*. Philadelphia: University of Pennsylvania Press.

Grossman, James R, Ann Durkin Keating, and Janice L. Reiff. 2004. *The Encyclopedia of Chicago*. Chicago: University of Chicago Press.

Hague, James D. 1870. *Mining Industry, by James D. Hague; With Geological Contributions by Clarence King*. Washington, DC: Government Printing Office.

Haines, Aubrey L. 1977. *The Yellowstone Story: A History of our First National Park*. Yellowstone National Park, WY: Yellowstone Library and Museum Association in cooperation with Colorado Associated University Press.

Hatley, Tom. 1991. *Dividing Paths: Cherokees and South Carolinians through the Era of Revolution*. New York: Oxford University Press.

Hedges, James Blaine. 1930. *Henry Villard and the Railways of the Northwest*. New Haven: Yale University Press.

Heidenreich, Conrad E. 1980. "Mapping the Great Lakes: The Period of Exploration, 1603–1700." *Cartographica* 17, no. 3: 32–63.

Heidenreich, Conrad E. and Edward H. Dahl. 1980. "The French Mapping of North America in the Seventeenth Century." *Map Collector* 13 (December): 2–11.

———. 1982. "The French Mapping of North America, 1700–1768." *Map Collector* 19 (June): 2–7.

Hine, Robert V. 1982. *In the Shadow of Frémont: Edward Kern and the Art of Exploration, 1845–1860*. Norman: University of Oklahoma Press.

Hine, Robert V., and John Mack Faragher. 2000. *The American West: A New Interpretive History*. New Haven and London: Yale University Press.

Hinsley, Curtis M. 1981. *The Smithsonian and the American Indian: Making Moral Anthropology in Victorian America*. Washington, DC: Smithsonian Institution Press.

Hirshson, Stanley P. 1967. *Grenville M. Dodge, Soldier, Politician, Railroad Pioneer*. Bloomington, Indiana University Press.

The History of Our Public Lands. c. 1882. "Justice Pamphlet No. 2." New York: Justice Publishing Co.

Holland, Robert A. 2005. *Chicago in Maps: 1612 to 2002*. New York: Rizzoli.

Hoover, John N. 1989. "A Man of Letters on the Frontier and His Library." In *A Guide to the John Mason Peck Collection of the St. Louis Mercantile Library Association*. St. Louis: St. Louis Mercantile Library Association.

Hoxie, Frederick E., ed. 1984. *A Final Promise: The Campaign to Assimilate the Indians, 1880–1920*. Lincoln, NE: University of Nebraska Press.

———. 1996. *Encyclopedia of North American Indians*. Boston: Houghton Mifflin Co.

Hyde, Anne F. 1990. *An American Vision: Far Western Landscape and American Culture, 1820–1920.* New York: New York University Press.

Jackson, Jack. 1995. *Manuscript Maps concerning the Gulf Coast, Texas, and the Southwest, 1519–1836.* Chicago: The Newberry Library.

Kagan, Richard L., and Fernando Marias. *Urban Images of the Hispanic World, 1493–1793.* New Haven, CT: Yale University Press, 2000.

Kain, Roger J. P., and Elizabeth Baigent. 1992. *The Cadastral Map in the Service of the State: A History of Property Mapping.* Chicago: University of Chicago Press.

Karrow, Robert W. 1984. "Made in Chicago: Maps and Atlases Printed in Chicago Before the Fire." In *Chicago Mapmakers: Studies in the Rise of the City's Map Trade*, ed. Michael P. Conzen, 73–76. Chicago: Chicago Historical Society for the Chicago Map Society, 1984.

——— 1986. "George M. Wheeler and the Geographical Surveys West of the 100th Meridian, 1869–1879." In *Exploration and Mapping of the American West: Selected Essays*, ed. Donna P. Koepp, 121–157. Chicago: Speculum Orbis Press, for the Map and Geography Roundtable of the American Library Association.

———. 2002a. "Heart of America: At Dead Center of the Union." In Paul E. Cohen, *Mapping the West: America's Westward Movement, 1524–1890,* 145–148. New York: Rizzoli.

———. 2002b. "The Wheeler Survey: George Wheeler's Sketch." In Paul E. Cohen, *Mapping the West: America's Westward Movement, 1524–1890,* 192–194. New York: Rizzoli.

Karrow, Robert W., and David Buisseret. 1984. *Gardens of Delight: Maps and Travel Accounts of Illinois and the Great Lakes from the Collection of Hermon Dunlap Smith: An Exhibition at The Newberry Library, 29 October–31 January 1985.* Chicago: Newberry Library.

Kaufman, Kevin. 1989. "Introduction," in *The Mapping of the Great Lakes in the Seventeenth Century.* Providence: John Carter Brown Library.

Khlebnikov, Kiril Timofeevich. 1994. *Notes of Russian America*, 2 vols., trans. Serge LeComte and Richard Pierce, ed. Richard Pierce. Kingston, Ontario: The Limestone Press.

Kohl, J. G. 1860. *Die beiden ältesten General-Karten von Amerika: Ausgeführt in den Jahren 1527 und 1529 auf Befehl Kaiser Karl's V. Im Besitz der Grossherzoglichen Bibliothek zu Weimar* [The two oldest general maps of America. Executed in the years 1527 and 1429 by command of Emperor Charles V. From the grand-ducal library in Weimar.] Weimar: Geographisches Institut, 1860.

Krogt, Peter van der. 1997–. *Koeman's Atlantes Neerlandici.* 'T Goy-Houten, the Netherlands: HES & De Graaf Publishers.

Koepp, Donna P., ed. 1986. *Exploration and Mapping of the American West: Selected Essays.* Chicago: Speculum Orbis Press, for the Map and Geography Roundtable of the American Library Association.

Lagarde, Lucie. 1989. "Le passage du Nord-Ouest et la Mer de l'ouest dans la cartographie française du 18è siècle, contribution à l'étude de l'oeuvre des Delisle et Buache." *Imago Mundi* 41:19–43.

Lahontan, Louis Armand de Lom d'Arce, Baron de. 1703. *New Voyages to North-America: Containing an Account of the Several Nations of that Vast Continent...* London: Printed for H. Bonwicke, T. Goodwin, M. Wotton, B. Tooke, and S. Manship, 1703.

Lamar, Howard R., ed. 1998. *The New Encyclopedia of the American West.* New Haven, CT: Yale University Press.

Lavender, David. 1992. *Let Me Be Free: The Nez Perce Tragedy.* New York: HarperCollins.

Lemmon, Alfred E., John T. Magill and Jason R. Wiese, eds. 2003. *Charting Louisiana: Five Hundred Years of Maps.* New Orleans: Historic New Orleans Collection.

Lewis, G. Malcolm. 1980. "Indian Maps." In *Old Trails and New Directions: Papers of the Third North American Fur Trade Conference*, edited by Carol M. Judd and Arthur J. Ray, 9–23. Toronto: University of Toronto Press.

———. 1986. "Indicators of Unacknowledged Assimilations from Amerindian 'Maps' on Euro-American Maps of North America: Some General Principles Arising From a Study of La Verendrye's Composite Map, 1728–9." *Imago Mundi* 38:9–34.

———. 1991. "La grande riviere et fleuve de l'ouest: the realities and reasons behind a major mistake in the eighteenth-century geography of North America." *Cartographica* 28:54–87.

———. 1987. "Misinterpretation of Amerindian Information as a Source of Error on Euro-American Maps." *Annals of the Association of American Geographers* 77:542–63.

———, ed. 1998a. *Cartographic Encounters: Perspectives on Native American Mapmaking and Map Use.* Chicago: University of Chicago Press.

———. 1998b. "Mapmaking, and Map Use by Native North Americans." In David Woodward and G. Malcolm Lewis, eds., *The History of Cartography*, Volume 2, Book 3, 51–182. Chicago and London: University of Chicago Press.

Lewis, Thomas A. 1993. *For King and Country: George Washington, the Early Years.* New York: John Wiley and Sons.

Lewty, Peter J. 1995. *Across the Columbia Plain: Railroad Expansion in the Pacific Northwest, 1885–1893.* Pullman, WA: Washington State University Press.

Library of Congress, Geography and Map Division, Reference and Bibliography Section. 1981. *Fire Insurance Maps in the Library of Congress: Plans of North American Cities and Towns Produced by the Sanborn Map Company: A Checklist.* Washington, DC: Library of Congress.

Luebke, Frederick C., Frances W. Kaye and Gary E. Moulton, eds. 1987. *Mapping the North American Plains.* Norman, OK: University of Oklahoma Press; Lincoln, NE: Center for Great Plains Studies, University of Nebraska–Lincoln.

Lutz, Alma. 1929. *Emma Willard, Daughter of Democracy.* Boston and New York: Houghton Mifflin Company.

Marshall, James. 1945. *Santa Fe, the Railroad that Built an Empire.* New York: Random House.

Martin, Albro. 1976. *James J. Hill and the Opening of the Northwest.* New York: Oxford University Press.

Martin, Lawrence. 1972. "John Disturnell's Map of the United Mexican States." In *A la Carte: Selected Papers on Maps and Atlases*, compiled by Walter W. Ristow. Washington, DC: Library of Congress.

Mattocks, Rev. John. 1867. "The Carver Centenary." *Collections of the Minnesota Historical Society* 2.

McDermott, John Francis. 1961. *Seth Eastman: Pictorial Historian of the Indian.* Norman: University of Oklahoma Press.

Mayer, Harold M. and Richard C. Wade. 1969. *Chicago: Growth of a Metropolis.* Chicago: University of Chicago Press.

Merrill, George P. 1927. *The First One Hundred Years of American Geology.* New Haven: Yale University Press.

Milner, Clyde A., II, Carol A. O'Connor, and Martha A. Sandweiss, eds. 1994. *The Oxford History of the American West.* New York and Oxford: Oxford University Press.

Modelski, Andrew M. 1975. *Railroad Maps of the United States.* Washington, DC: Library of Congress.

———. 1984. *Railroad Maps of North America: The First Hundred Years.* Washington, DC: Library of Congress.

Mollat du Jourdin, Michel, and Monique de La Roncière. 1984. *Sea Charts of the Early Explorers, 13th to 17th Century.* New York: Thames and Hudson.

Morgan, Dale L. 1953. *Jedediah Smith and the Opening of the West.* New York: Bobbs-Merrill Co.

Morgan, Dale L. and Carl I. Wheat. 1954. *Jedediah Smith and his Maps of the American West.* San Francisco: California Historical Society.

Morris, John W., Charles R. Goins, and Edwin C. MacReynolds. 1976. *Historical Atlas of Oklahoma.* Norman: University of Oklahoma Press.

Moulton, Gary E., ed. 1983. *Atlas of the Lewis & Clark Expedition.* Lincoln: University of Nebraska Press.

———, ed. 2003. *The Lewis and Clark Journals: An American Epic of Discovery.* Lincoln: University of Nebraska Press.

Mumey, Nolie. 1948. *Pioneer Denver, Including Scenes of Central City, Colorado City and Nevada City.* Dillingham Lithographs (1861–1863). Denver: Artcraft Press.

Mundy, Barbara E. 1996. *The Mapping of New Spain: Indigenous Cartography and the Maps of the Relaciones Geográficas.* Chicago: University of Chicago Press.

Musich, Jerry. 2006. "Mapping a Transcontinental Nation: Nineteenth and Early Twentieth-Century Rail Travel Cartography." In *Cartographies of Travel and Navigation*, ed. James R. Akerman. Chicago: University of Chicago Press.

Nabokov, Peter and Robert Easton. 1989. *Native American Architecture.* New York: Oxford University Press.

Nebenzahl, Kenneth. 1986. "Mapping the Trans-Mississippi West: Annotated Selections." In *Exploration and Mapping of the American West: Selected Essays* ed. Donna P. Koepp, 1–23. Chicago: Speculum Orbis Press, for the Map and Geography Roundtable of the American Library Association.

———. 1990. *Atlas of Columbus and the Great Discoveries*. Rand McNally.

Nelson, George. 1998. *The Alamo: An Illustrated History*. Dry Frio Canyon, TX: Aldine Press.

Nordholt, Jan Willem Schulte. 1995. *The Myth of the West: America as the Last Empire*. Grand Rapids, MI.

O'Connor, Richard. 1973. *Iron Wheels and Broken Men: The Railroad Barons and the Plunder of the West*. New York: Putnam.

Ohl, Darryl L. 1980. *Directory of the City of Davenport for 1856*. Davenport: Scott County Iowa Genealogical Society.

Ortiz, Alfonso, ed. 1979. *Handbook of North American Indians*, Volume 9: Southwest, general ed. William Sturtevant. Washington, DC: Smithsonian Institution.

Overton, Richard C. 1941. *Burlington West: A Colonization History of the Burlington Railroad*. Cambridge: Harvard University Press.

Pagden, Anthony. 1998. *Lords of All the World: Ideologies of Empire in Spain, Britain, and France, c. 1500–c. 1800*. New Haven, CT: Yale University Press.

Parke-Bernet Galleries. 1969. *The Original Manuscript Map of the Military District, Kansas and the Territories (1866) by Maj. Gen. Grenville Mellen Dodge*. Auction catalog, sale no. 2934A. New York: Parke-Bernet Galleries.

Parker, John, ed. 1976. *The Journals of Jonathan Carver and Related Documents*, 1766–1770. St. Paul: Minnesota Historical Society Press.

———. 1986. "New Light on Jonathan Carver." *The American Magazine and Historical Chronicle* 2 (Spring/Summer): 4–17.

Patton, Jeffrey C. 1999. "The American School Atlas: 1794–1900." *Cartographic Perspectives* 33 (Spring): 4–32.

Pedley, Mary Sponberg. 2005. *The Commerce of Cartography: Making and Marketing Maps in Eighteenth-Century France and England*. Chicago: University of Chicago Press.

Pelletier, Monique. 1983. "Exploration and Colonisation of Louisiana." *Map Collector* 24 (September):10–14.

Perrigo, Lynn I. 1934. *The Little Kingdom: A Record Chiefly of Central City in the Early Days*. Boulder, CO.

Peters, Cynthia H. 1984. "Rand, McNally in the Nineteenth Century: Reaching for a National Market." In *Chicago Mapmakers: Studies in the Rise of the City's Map Trade*, ed. Michael P. Conzen, 64–72. Chicago: Chicago Historical Society for the Chicago Map Society.

Pfeiffer, David A. 2004. "Bridging the Mississippi: The Railroads and Steamboats Clash at the Rock Island Bridge." *Prologue Magazine* 36 (Summer).

Pierce, Richard A. 1986. *Builders of Alaska: The Russian Governors, 1818–1867*. Kingston, Ontario: Limestone Press.

———, ed. 1990a. *Russia in North America: Proceedings of the 2nd International Conference on Russian America, Sitka, Alaska, August 19–22, 1987*. Kingston, Ontario: The Limestone Press.

———. 1990b. *Russian America: A Biographical Dictionary*. Kingston, Ontario: Limestone Press.

Powell, John Wesley. 1879. *Report on the Lands of the Arid Region of the United States*. Washington, DC: Government Printing Office.

———. 1895. *The Canyons of the Colorado*. Meadville, PA: Flood and Vincent.

Preuss, Charles. 1958. *Exploring with Frémont*, ed. and trans., Erwin G. and Elizabeth K. Gudde. Norman, OK: Oklahoma University Press.

Pritchard, Margaret Beck, and Henry G. Taliaferro. 2002. *Degrees of Latitude: Mapping Colonial America*. New York: Henry N. Abrams, Inc., for the Colonial Williamsburg Foundation.

Prucha, Francis Paul. 1984. *The Great Father: The U.S. Government and the American Indian*. Lincoln: University of Nebraska Press.

———, ed. 1990. *Documents of the United States Indian Policy*. Lincoln: University of Nebraska Press.

Quaife, Milo M. 1920. "Jonathan Carver and the Carver Grant." *The Mississippi Valley Historical Review* 7, no. 1 (June): 3–25.

Rabbitt, Mary C. 1979–. *Minerals, Lands, and Geology for the Common Defence and General Welfare: A History of Public Lands, Federal Science and Mapping Policy, and Development of Mineral Resources in the United States*. Vol. 1. Before 1879; vol. 2. 1879–1903; vol. 3. 1904–1939. Washington: U.S. Geological Survey.

Rebert, Paula. 2001. *La Gran Linea: Mapping the United States-Mexico Boundary, 1849–1857*. Austin: University of Texas Press.

Reed, Susan M. 1915. "British Cartography of the Mississippi Valley in the Eighteenth Century." *Mississippi Valley Historical Review* 2:213–224.

Rees, William. 1854. *The Mississippi Bridge Cities, Davenport, Rock Island and Moline*. Rock Island: H. W. Porter and Brother.

Renz, Louis Tuck. 1980. *History of the Northern Pacific Railroad*. Fairfield, WA: Ye Galleon Press.

Reps, John W. 1965. *The Making of Urban America: A History of City Planning in the United States*. Princeton, NJ: Princeton University Press.

———. 1984. *Views and Viewmakers of Urban America*. Columbia: University of Missouri Press.

Richter, Daniel K. 2001. *Facing East from Indian Country: A Native History of Early America*. Cambridge: Harvard University Press.

Ristow, Walter W. 1968. "United States Fire Insurance and Underwriters Maps, 1852–1968." *Quarterly Journal of the Library of Congress* 25:194–217.

———, ed. 1972. *A La Carte: Selected Papers on Maps and Atlases*. Washington, DC: Library of Congress.

———. 1977. *Maps for an Emerging Nation: Commercial Cartography in Nineteenth-Century America*. Washington, DC: Library of Congress.

———. 1985. *American Maps and Mapmakers: Commercial Cartography in the Nineteenth Century*. Detroit, MI: Wayne State University Press.

Robbins, William G. 1994. *Colony and Empire: The Capitalist Transformation of the American West*. Lawrence: University Press of Kansas.

Rowley, Gwyn. 1984. "An Introduction to British Fire Insurance Plans." *Map Collector* 29 (December):14–19.

Runte, Alfred. 1974. "Pragmatic Alliance: Western Railroads and the National Parks." *National Parks* 48 (April): 14–24.

———. 1994. *Trains of Discovery: Western Railroads and the National Parks*. Niwot, Colorado: Roberts Rinehart Publishers.

Santos, Richard G. 1981. *Aguayo Expedition into Texas, 1721: An Annotated Translation of the Five Versions of the Diary Kept by Br. Juan Antonio de la Pena*. Austin, TX: Jenkins Publishing Co.

Schoelwer, Susan Prendergast. 1985. *Alamo Images: Changing Perceptions of a Texas Experience*. Dallas: DeGolyer Library and Southern Methodist University Press.

Schulten, Susan. 2001. *The Geographical Imagination in America, 1880–1950*. Chicago: University of Chicago Press.

———. 2007. "Mapping American History." In *Maps: Finding Our Place in the World*, ed. James R. Akerman and Robert W. Karrow, Jr. Chicago and London: University of Chicago Press.

Schwartz, Seymour I. 1994. *The French and Indian War, 1754–1763: The Imperial Struggle for North America*. New York: Simon & Schuster.

Schwartz, Seymour I. and Ralph E. Ehrenberg. 1980. *The Mapping of North America*. New York: H.N. Abrams.

Schwartz, Seymour I. and Henry Taliaferro. 1984. "A Newly Discovered First State of a Foundation Map 'L'Amerique Septentrionale'." *Map Collector* 26 (March): 2-6.

Selmer, Marsha L. 1984. "Rufus Blanchard: Early Chicago Map Publisher." In *Chicago Mapmakers: Studies in the Rise of the City's Map Trade*, ed. Michael P. Conzen. Chicago: Chicago Historical Society for the Chicago Map Society, pp. 23–31.

Shannon, Timothy J. 2002. *Indians and Colonists at the Crossroads of Empire*. Ithaca: Cornell University Press.

Shirley, Rodney. 1983. *The Mapping of the World: Early Printed World Maps, 1472–1700*. Holland Press Cartographica; v. 9. London: Holland Press.

Short, John R. 2001. *Representing the Republic: Mapping the United States, 1600–1900.* London: Reaktion.

Simpson, James H. 1964. *Navaho Expedition: Journal of a Military Reconnaissance from Santa Fe, New Mexico, to the Navaho Country made in 1849.* Edited and annotated by Frank McNitt. Norman: University of Oklahoma Press.

Skelton, R. A. 1965. Introduction to Gerard de Jode, *Speculum Orbis Terrarum.* Facsimile edition. Amsterdam: Theatrum Orbis Terrarum.

Smith, Barbara S., and Redmond J. Barnette, eds. 1990. *Russian America: The Forgotten Frontier.* Tacoma: Washington State Historical Society.

Smith, Henry Nash. 1973. *Virgin Land: The American West as Symbol and Myth.* Cambridge: Harvard University Press; orig. 1950.

Stegner, Wallace. 1953. *Beyond the Hundredth Meridian: John Wesley Powell and the Second Opening of the West.* Boston: Houghton Mifflin.

Stevenson, E. L. 1909. *Early Spanish Cartography of the New World.* Worcester: American Antiquarian Society.

Stover, John F. 1978. *Iron Road to the West: American Railroads in the 1850s.* New York: Columbia University Press.

Swinton, John R. 1937. "William Henry Holmes." *National Academy of Science Biographical Memoirs,* vol. XVII. Washington DC: National Academy of Sciences.

Teben'kov, M. D. 1981. *Atlas of the Northwest Coasts of America from Bering Straits to Cape Corrientes, and the Aleutian Islands, with several sheets on the Northeast Coast of Asia,* trans. and ed. R. A. Pierce. Kingston, Ontario: The Limestone Press.

Thompson, M. M. 1981. *Maps for America: Cartographic Products of the U.S. Geological Survey.* Washington, DC: Government Printing Office.

Thrower, Norman W. 1986. "Mapping the United States–Mexico Borderlands: An Overview." In *Exploration and Mapping of the American West: Selected Essays,* ed. Donna P. Koepp, 159–169. Chicago: Speculum Orbis Press, for the Map and Geography Roundtable of the American Library Association.

Thurman, Mel. 1989. "Warren, Dodge and Later Nineteenth-Century Army Maps of the American West." *Mapline* 53 (March): 1-4.

Time-Life Books staff. 1976. *The Spanish West.* New York.

Trachtenberg, Alan. 1989. *Reading American Photographs: Images as History.* New York: Hill and Wang.

Tyler, Ron, et al. 1987. *American Frontier Life: Early Western Painting and Prints.* New York: Abbeville Press.

United States Senate. 1890. "Letter from the Secretary of the Interior, transmitting, In response to the Senate resolution of March 10, 1890, the compilation concerning the legal status of the Indians in Indian Territory." In U.S. 51st Congress, 1st session, 1889–1890. Senate Executive Document 78, 1–31.

U.S. House Committee on Public Lands 1825. *Indian Grant to Captain Carver,* 18th Cong., 2nd sess., 1825, S. Rep. 444, 82-89.

U.S. vs. Burlington & Missouri Railroad Co. 1878. 98 U. S. 334.

Vance, James E. 1995. *The North American Railroad: Its Origin, Evolution, and Geography.* Baltimore: Johns Hopkins University Press.

Verner, Coolie, and Basil Stuart Stubbs. 1979. *The Northpart of America.* Academic Press Canada.

Vigneres, L. A. 1962. "The Cartographer Diogo Ribeiro." *Imago Mundi* 16: 73–83.

Villard, Henry. 1904. *Memoirs of Henry Villard, Journalist and Financier, 1835–1900...* Boston and New York: Houghton, Mifflin and Co.

Wagner, Henry R. 1928. "Spanish Voyages to the Northwest Coast of America during the Sixteenth Century." San Francisco; California Historical Society.

———. 1931. "The Manuscript Atlases of Battista Agnese." *Papers of the Bibliographical Society of America* 25: 1–110.

———. 1937a. "Cartography of the Northwest Coast of America to the Year 1800." 2 vols., Berkeley: University of California Press.

———. 1937b. *The Spanish Southwest, 1542–1794,* 2 vols. Albuquerque.

Walker, Deward E., Jr. 1998. *Handbook of North American Indians,* Volume 12: Plateau, general ed. William Sturtevant. Washington, DC: Smithsonian Institution.

Wallis, Helen. 1997. "Sixteenth-Century Maritime Manuscript Atlases for Special Presentation." In *Images of the World: The Atlas Through History,* ed. John A. Wolter and Ronald E. Grim, 3–29. Washington, DC: Library of Congress.

Warhus, Mark. 1997. *Another America: Native American Maps and the History of Our Land.* New York: St. Martin's Press.

Warren, G. K. 1856. Essay on early maps of Western America, in *Pacific Railroad Reports,* vol. 9, Part 2, Washington, DC.

Washburn, Wilcomb E. 1988. *Handbook of North American Indians,* Volume 4: History of Indian-White Relations, general ed. William Sturtevant. Washington, DC: Smithsonian Institution.

Weber, David J. 1982. *The Mexican Frontier, 1821–1846: The American Southwest Under Mexico.* Albuquerque: University of New Mexico Press.

Weber, G. A. 1923. *The Coast and Geodetic Survey: Its History, Activities and Organization.* Service Monograph of the U.S. Government, no. 16. Baltimore: Johns Hopkins University Press.

Wheat, Carl I. 1942. *The Maps of the California Gold Region, 1848–1857: A Biblio-cartography of an Important Decade.* San Francisco: Grabhorn Press.

———. 1949. *Books of the California Gold Rush. A Centennial Selection.* San Francisco: Colt Press.

———. 1957–1963. *Mapping the Transmississippi West,* 5 vols. San Francisco: Institute for Historical Cartography.

White, Richard. 1983. *The Roots of Dependency: Subsistence, Environment, and Social Change among the Choctaw, Pawnees, and Navajos.* Lincoln: University of Nebraska Press.

———. 1991a. *"It's Your Misfortune and None of My Own": A History of the American West.* Norman and London: University of Oklahoma Press.

———. 1991b. *The Middle Ground: Indians, Republics, and Empires in the Great Lakes Region, 1680–1815.* New York: Cambridge University Press.

———. 1994. "Frederick Jackson Turner and Buffalo Bill." In *The Frontier in American Culture,* ed. James R. Grossman, 7–65. Berkeley: University of California Press.

Wilkie, Franc B. 1858. *Davenport, Past and Present.* Davenport: Luse, Lane and Co.

Williams, Kenny J. 1980. *Prairie Voices: A Literary History of Chicago from the Frontier to 1893.* Nashville: Townsend Press.

Winsor, Justin. 1904. *The Kohl Collection (now in the Library of Congress) of Maps Relating to America.* Washington, DC: Government Printing Office.

Winther, Oscar Osburn. 1964. *The Transportation Frontier: Trans-Mississippi West, 1865–1890.* Washington, DC: Government Printing Office.

Wolter, John A. 1981. "Johann Georg Kohl and America." *Map Collector* 17 (December): 10–14.

Wolter, John A., and Ronald E. Grim, eds. 1997. *Images of the World: The Atlas Through History.* Washington, DC: Library of Congress.

Woodward, David. 1977. *The All-American Map: Wax Engraving and its Influence on Cartography.* Chicago: University of Chicago Press.

———. 1997. "Italian Composite Atlases of the Sixteenth Century." In *Images of the World: The Atlas Through History,* ed. John A. Wolter and Ronald E. Grim, 51–70. Washington, DC: Library of Congress.

Worms, Laurence. 1993. "Thomas Kitchin's Journey of Life: Hydrographer to George III, Mapmaker and Engraver." Parts I and II in *Map Collector* 62:2–8; 63:14–20.

Worster, Donald. 2001. *A River Running West: The Life of John Wesley Powell.* Oxford: Oxford University Press.

Wright, William [Dan De Quille, pseud.]. 1877. *History of the Big Bonanza: An Authentic Account of the Discovery, History, and Working of the World Renowned Comstock Silver Lode of Nevada...* Hartford, CT: American Pub. Co.; San Francisco, CA: A.L. Bancroft & Co.

MOHAWK windpower

This book is printed on Mohawk Via Bright White and Mohawk Via Natural, which contain 30% postconsumer waste fiber and are manufactured with windpower.